SECRE+S

OF

MEN+AL MA+H

SECRE+S

OF

MEN+AL MA+H

The Mathemagician's Guide to Lightning Calculation
and Amazing Math Tricks

Arthur Benjamin
and **Michael Shermer**

THREE RIVERS PRESS • NEW YORK

Published in the United States by Three Rivers Press, an imprint of the Crown
Publishing Group, a division of Random House, Inc., New York.
www.crownpublishing.com

Originally published in different form as *Mathemagics* by Lowell House,
Los Angeles, in 1993.

Three Rivers Press and the Tugboat design are registered trademarks of Random
House, Inc.

Library of Congress Cataloging-in-Publication Data

Benjamin, Arthur.
  Secrets of mental math : the mathemagician's guide to lightning calculation and
amazing math tricks / Arthur Benjamin and Michael Shermer.— 1st ed.
      p. cm.
  Includes bibliographical references and index.
  1. Mental arithmetic—Study and teaching.   2. Magic tricks in mathematics
education.   3. Mental calculators.   I. Shermer, Michael.   II. Title.
  QA111.B44   2006
  510—dc22

2005037289

ISBN-13: 978-0-307-33840-2
ISBN-10: 0-307-33840-1 (alk. paper)

Printed in the United States of America

Design by Ruth Lee-Mui

20 19 18 17 16 15 14 13 12 11

First Edition

I dedicate this book to my wife, Deena,
and daughters, Laurel and Ariel.

—Arthur Benjamin

My dedication is to my wife, Kim,
for being my most trusted confidante
and personal counselor.

—Michael Shermer

Acknowledgments

The authors wish to thank Steve Ross and Katie McHugh at Random House for their support of this book. Special thanks to Natalya St. Clair for typesetting the initial draft, which was partly supported by a grant from the Mellon Foundation.

Arthur Benjamin especially wants to acknowledge those who inspired him to become both a mathematician and a magician—cognitive psychologist William G. Chase, magicians Paul Gertner and James Randi, and mathematicians Alan J. Goldman and Edward R. Scheinerman. Finally, thanks to all of my colleagues and students at Harvey Mudd College, and to my wife, Deena, and daughters, Laurel and Ariel, for constant inspiration.

Contents

Foreword

by Bill Nye (the Science Guy®)

I like to think about the first humans, the people who came up with the idea to count things. They must have noticed right away that figuring on your fingertips works great. Perhaps Og (a typical ancient cave guy) or one of his pals or associates said, "There are one, two, three, four, five of us here, so we need five pieces of fruit." Later, "Hey, look," someone must have said (or grunted), "you can count the number of people at the campfire, the number of birds on a tree, stones in a row, logs for a fire, or grapes in a bunch, just with your fingers." It was a great start. It's probably also how *you* came to first know numbers.

You've probably heard that math is the language of science, or the language of Nature is mathematics. Well, it's true. The more we understand the universe, the more we discover its mathematical connections. Flowers have spirals that line up with a special sequence of numbers (called Fibonacci numbers) that you can understand and generate yourself. Seashells form in perfect mathematical curves (logarithmic spirals) that come from a chemical balance. Star clusters tug on one another in a mathematical dance that we can observe and understand from millions and even billions of kilometers away.

We have spent centuries discovering the mathematical nature of Nature. With each discovery, someone had to go through the math and make sure the numbers were right. Well, *Secrets of Mental Math* can help you handle all kinds of numbers. You'll get comfortable with calculations in a way that will let you know some of Nature's numerical secrets, and who knows where that might take you?

As you get to know numbers, the answer really is at your fingertips. That's not a joke, because that's where it all begins. Almost everyone has ten fingers, so our system of mathematics started with 1 and went to 10. In fact, we call both our numbers and our fingers "digits." Coincidence? Hardly. Pretty soon, though, our ancestors ran out of fingers. The same thing has probably happened to you. But we can't just ignore those big numbers and (this is a joke) throw up our hands.

We need numbers—they're part of our lives every day, and in ways we typically don't even notice. Think about a conversation you had with a friend. To call, you needed a phone number, and the time you spent on the phone was measured in numbers of hours and minutes. Every date in history, including an important one like your birthday, is reckoned with numbers. We even use numbers to represent ideas that have nothing to do with counting. What's your 20? (I.e., Where are you? From the old police "10" codes, like 10-4 for "yes.") What's the 411 on that gal? (I.e., What's her background; is she dating anyone? From the number for telephone information.) People describe one another in numbers representing height and weight. And, of course, we all like to know how much money we have or how much something costs in numbers: dollars, pesos, yuan, rupees, krona, euros, or yen. Additionally (another joke), this book has a time-saving section on remembering numbers—and large numbers of numbers.

If, for some reason, you're not crazy about math, read a little further. Of course I, as the Science Guy, hope you do like math.

Well, actually, I hope you *love* math. But no matter how you feel about math, hatred or love, I'd bet that you often find yourself just wanting to know the answer right away, without having to write down everything carefully and work slowly and diligently—or without even having to stop and grab a calculator. You want the answer, as we say, "as if by magic." It turns out that you can solve or work many, many math problems almost magically. This book will show you how.

What makes any kind of magic so intriguing and fun is that the audience seldom knows how the trick is performed. "How did she do that . . . ?" "I don't know, but it's cool." If you have an audience, the tricks and shortcuts in *Secrets of Mental Math* are a lot like magic. The audience seldom knows how a trick is performed; they just appreciate it. Notice, though, that in magic, it's hardly worth doing if no one is watching. But with *Secrets,* knowing how it works doesn't subtract from the fun (or pun). When arithmetic is easy, you don't get bogged down in the calculating; you can concentrate on the wonderful nature of numbers. After all, math runs the universe.

Dr. Benjamin got into this business of lightning-fast calculating just for fun. We have to figure he impressed his teachers and classmates. Magicians might make some in their audience think that they have supernatural powers. Mathemagicians, at first, give the impression that they're geniuses. Getting people to notice what you're doing is an old part of sharing ideas. If they're impressed, they'll probably listen to what you have to say. So try some "mathemagics." You may impress your friends, all right. But you'll also find yourself performing just for yourself. You'll find you're able to do problems that you didn't think you could. You'll be impressed . . . *with yourself.*

Now, counting on your fingers is one thing (one finger's worth). But have you ever found yourself counting out loud or whispering or making other sounds while you calculate? It almost always makes math easier. The problem, though, is that other people think you're a little odd . . . not even (more math humor). Well, in *Secrets of Mental Math*, Dr. Benjamin helps you learn to use that "out-loud" feature of the way your brain works to do math problems more easily, faster, and more accurately (which is surprising), all while your brain is thinking away—almost as if you're thinking out loud.

You'll learn to move through math problems the same way we read in English, left to right. You'll learn to handle big problems fast with good guesses, actually great guesses, within a percent or so. You will learn to do arithmetic fast; that way you can spend your time thinking about what the numbers mean. Og wondered, "Do we have enough fruit for each person sitting around the fire? If not, there might be trouble." Now you might wonder, "Is there enough space on this computer to keep track of my music files . . . or my bank account? If not, there might be trouble."

There's more to *Secrets* than just figuring. You can learn to take a day, month, and year, then compute what day of the week it was or will be. It's fantastic, almost magical, to be able to tell someone what day of the week she or he was born. But, it's really something to be able to figure that the United States had its first big Fourth of July on a Thursday in 1776. April 15, 1912, the day the *Titanic* sank, was a Monday. The first human to walk on the moon set foot there on July 20, 1969, a Sunday. You'll probably never forget that the United States was attacked by terrorists on September 11, 2001. With *Secrets of Mental Math*, you'll always be able to show it was a Tuesday.

There are relationships in Nature that numbers describe better than any other way we know. There are simple numbers that you can count on your hands: one, two, three, and on up. But there are also an infinite number of numbers in between. There are fractions. There are numbers that never end. They get as big as you want and so small that they're hard to imagine. You can know them. With *Secrets of Mental Math*, you can have even these in-between numbers come so quickly to your mind that you'll have a bit more space in your brain to think about why our world works this way. One way or another, this book will help you see that in Nature, it all adds up.

Foreword
by James Randi

Mathematics is a wonderful, elegant, and exceedingly useful language. It has its own vocabulary and syntax, its own verbs, nouns, and modifiers, and its own dialects and patois. It is used brilliantly by some, poorly by others. Some of us fear to pursue its more esoteric uses, while a few of us wield it like a sword to attack and conquer income tax forms or masses of data that resist the less courageous. This book does not guarantee to turn you into a Leibniz, or put you on stage as a Professor Algebra, but it will, I hope, bring you a new, exciting, and even entertaining view of what can be done with that wonderful invention—numbers.

We all think we know enough about arithmetic to get by, and we certainly feel no guilt about resorting to the handy pocket calculator that has become so much a part of our lives. But, just as photography may blind us to the beauty of a Vermeer painting, or an electronic keyboard may make us forget the magnificence of a Horowitz sonata, too much reliance on technology can deny us the pleasures that you will find in these pages.

I remember the delight I experienced as a child when I was shown that I could multiply by 25 merely by adding two 0s to my number and dividing by 4. Casting out 9s to check multiplication came next, and when I found out about cross-multiplying I was hooked and became, for a short while, a generally unbearable math nut. Immunizations against such afflictions are not available. You have to recover all by yourself. Beware!

This is a fun book. You wouldn't have it in your hands right now if you didn't have some interest either in improving your math skills or in satisfying a curiosity about this fascinating subject. As with all such instruction books, you may retain and use only a certain percentage of the varied tricks and methods described here, but that alone will make it worth the investment of your time.

I know both the authors rather well. Art Benjamin is not only one of those whiz kids we used to groan about in school but also has been known to tread the boards at the Magic Castle in Hollywood, performing demonstrations of his skill, and on one occasion he traveled to Tokyo, Japan, to pit his math skills against a lady savant on live television. Michael Shermer, with his specialized knowledge of science, has an excellent overview of practical applications of math as it is used in the real world.

If this is your first exposure to this kind of good math stuff, I envy you. You'll discover, as you come upon each delicious new way to attack numbers, that you missed something in school. Mathematics, particularly arithmetic, is a powerful and dependable tool for day-to-day use that enables us to handle our complicated lives with more assurance and accuracy. Let Art and Michael show you how to round a few of the corners and cut through some of the traffic. Remember these words of Dr. Samuel Johnson, an eminently practical soul in all respects: "Arithemetical inquiries give entertainment in solitude by the practice, and reputation in public by the effect."

Above all, enjoy the book. Let it entertain you, and have fun with it. That, with the occasional good deed, a slice of pizza (no anchovies!), and a selection of good friends is about all you can ask of life. Well, almost all. Maybe a Ferrari . . .

Prologue
by Michael Shermer

My good friend Dr. Arthur Benjamin, mathematics professor at Harvey Mudd College in Claremont, California, takes the stage to a round of applause at the Magic Castle, a celebrated magic club in Hollywood, where he is about to perform "mathemagics," or what he calls the art of rapid mental calculation. Art appears nothing like a mathematics professor from a prestigious college. Astonishingly quick-witted, he looks at home with the rest of the young magicians playing at the Castle—which he is.

What makes Art so special is that he can play in front of any group, including professional mathematicians and magicians, because he can do something that almost no one else can. Art Benjamin can add, subtract, multiply, and divide numbers in his head faster than most people can with a calculator. He can square two-digit, three-digit, and four-digit numbers, as well as find square roots and cube roots, without writing anything down on paper. And he can teach you how to perform your own mathematical magic.

Traditionally, magicians refuse to disclose how they perform their tricks. If they did, everyone would know how they are done and the mystery and fascination of magic would be lost. But Art wants to get people excited about math. And he knows that one of the best ways to do so is to let you and other readers in on his secrets of "math genius." With these skills, almost any-one can do what Art Benjamin does every time he gets on stage to perform his magic.

This particular night at the Magic Castle, Art begins by ask-ing if anyone in the audience has a calculator. A group of engi-neers raise their hands and join Art on the stage. Offering to test their calculators to make sure they work, Art asks a member of the audience to call out a two-digit number. "Fifty-seven," shouts one. Art points to another who yells out, "Twenty-three."

Directing his attention to those on stage, Art tells them: "Multiply 57 by 23 on the calculator and make sure you get 1311 or the calculators are not working correctly." Art waits patiently while the volunteers finish inputting the numbers. As each participant indicates his calculator reads 1311, the audi-ence lets out a collective gasp. The amazing Art has beaten the calculators at their own game!

Art next informs the audience that he will square four two-digit numbers faster than his button-pushers on stage can square them on their calculators. The audience asks him to square the numbers 24, 38, 67, and 97. Then, in large, bold writing for everyone to see, Art writes: 576, 1444, 4489, 9409. Art turns to his engineer volunteers, each of whom is computing a two-digit square, and asks them to call out their answers. Their response triggers gasps and then applause from the audi-ence: "576, 1444, 4489, 9409." The woman next to me sits with her mouth open in amazement.

Art then offers to square three-digit numbers without even writing down the answer. "Five hundred and seventy-two," a gentleman calls out. Art's reply comes less than a second later: "572 squared is 327,184." He immediately points to another member of the audience, who yells, "389," followed by Art's unblinking response: "389 squared will give you 151,321."

Someone else blurts out, "262." "That'll give you 68,644." Sensing he delayed just an instant on that last one, he promises to make up for it on the next number. The challenge comes—991. With no pause, Art squares the number, "982,081." Several more three-digit numbers are given and Art responds perfectly. Members of the audience shake their heads in disbelief.

With the audience in the palm of his hand, Art now declares that he will attempt to square a four-digit number. A woman calls out, "1,036," and Art instantly responds, "That's 1,073,296." The audience laughs and Art explains, "No, no, that's much too easy a number. I'm not supposed to beat the calculators on these. Let's try another one." A man offers a challenging 2,843. Pausing briefly between digits, Art responds: "Let's see, the square of that should be 8 million . . . 82 thousand . . . 649." He is right, of course, and the audience roars their approval, as loudly as they did for the previous magician who sawed a woman in half and made a dog disappear.

It is the same everywhere Art Benjamin goes, whether it is a high school auditorium, a college classroom, a professional conference, the Magic Castle, or a television studio. Professor Benjamin has performed his special brand of magic live all over the country and on numerous television talk shows. He has been the subject of investigation by a cognitive psychologist at Carnegie Mellon University and is featured in a scholarly book by Steven Smith called *The Great Mental Calculators: The Psychology, Methods, and Lives of Calculating Prodigies, Past and Present.* Art was born in Cleveland on March 19, 1961 (which he calculates was a Sunday, a skill he will teach you in Chapter 9). A hyperactive child, Art drove his teachers mad with his classroom antics, which included correcting the mathematical mistakes they occasionally made. Throughout this book when teaching you his mathematical secrets, Art recalls when and where he learned these skills, so I will save the fascinating stories for him to tell you.

Art Benjamin is an extraordinary individual with an extraordinary program to teach you rapid mental calculation. I offer these claims without hesitation, and ask only that you remember this does not come from a couple of guys promising miracles if you will only call our 800 hotline. Art and I are both credentialed in the most conservative of academic professions—Art in mathematics and I, myself, in the history of science—and we would never risk career embarrassment (or worse) by making such powerful claims if they were not true. To put it simply, this stuff works, and virtually everyone can do it because this art of "math genius" is a learned skill. So you can look forward to improving your math skills, impressing your friends, enhancing your memory, and, most of all, having fun!

Introduction

Ever since I was a child, I have loved playing with numbers, and in this book I hope to share my passion with you. I have always found numbers to have a certain *magical* appeal and spent countless hours entertaining myself and others with their beautiful properties. As a teenager, I performed as a magician, and subsequently combined my loves of math and magic into a full-length show, called Mathemagics, where I would demonstrate and explain the secrets of rapid mental calculation to audiences of all ages.

Since earning my PhD, I have taught mathematics at Harvey Mudd College, and I still enjoy sharing the joy of numbers with children and adults throughout the world. In this book, I will share all of my secrets for doing math in your head, quickly and easily. (I realize that magicians are not supposed to reveal their secrets, but *mathemagicians* have a different code of ethics. Mathematics should be awe inspiring, not mysterious.)

What will you learn from this book? You will learn to do math in your head faster than you ever thought possible. After a little practice, you will dramatically improve your memory for numbers. You will learn feats of mind that will impress your friends, colleagues, and teachers. But you will also learn to view math as an activity that can actually be *fun*.

Too often, math is taught as a set of rigid rules, leaving little room for *creative* thinking. But as you will learn from *Secrets*, there are often several ways to solve the same problem. Large problems can be broken down into smaller, more manageable components. We look for special features to make our problems easier to solve. These strike me as being valuable life lessons that we can use in approaching all kinds of problems, mathematical and otherwise.

"But isn't math talent something that you are born with?" I get this question all the time. Many people are convinced that lightning calculators are prodigiously gifted. Maybe I was born with some curiosity about how things work, whether it be a math problem or a magic trick. But I am convinced, based on many years of teaching experience, that rapid math is a skill that anyone can learn. And like any worthwhile skill, it takes practice and dedication if you wish to become an expert. But to achieve these results efficiently, it is important that you practice the *right* way. Let me show you the way!

Mathemagically,
Dr. Arthur Benjamin
Claremont, California

Quick Tricks:
Easy (and Impressive) Calculations

In the pages that follow, you will learn to do math in your head faster than you ever thought possible. After practicing the methods in this book for just a little while, your ability to work with numbers will increase dramatically. With even more practice, you will be able to perform many calculations faster than someone using a calculator. But in this chapter, my goal is to teach you some easy yet impressive calculations you can learn to do immediately. We'll save some of the more serious stuff for later.

INSTANT MULTIPLICATION

Let's begin with one of my favorite feats of mental math—how to multiply, in your head, any two-digit number by eleven. It's very easy once you know the secret. Consider the problem:

$$32 \times 11$$

To solve this problem, simply add the digits, $3 + 2 = \underline{5}$, put the 5 between the 3 and the 2, and there is your answer:

3<u>5</u>2

What could be easier? Now you try:

53 × 11

Since 5 + 3 = 8, your answer is simply

583

One more. Without looking at the answer or writing anything down, what is

81 × 11?

Did you get 891? Congratulations!

Now before you get too excited, I have shown you only half of what you need to know. Suppose the problem is

85 × 11

Although 8 + 5 = <u>13</u>, the answer is NOT 81<u>3</u>5!

As before, the <u>3</u> goes in between the numbers, but the <u>1</u> needs to be added to the 8 to get the correct answer:

9<u>3</u>5

Think of the problem this way:

1

<u>835</u>

935

Here is another example. Try 57 × 11.
Since 5 + 7 = 12, the answer is

I

527

627

Okay, now it's your turn. As fast as you can, what is

77 × 11?

If you got the answer 847, then give yourself a pat on the back. You are on your way to becoming a mathemagician.

Now, I know from experience that if you tell a friend or teacher that you can multiply, in your head, any two-digit number by eleven, it won't be long before they ask you to do 99 × 11. Let's do that one now, so we are ready for it.

Since 9 + 9 = 18, the answer is:

I

989

1089

Okay, take a moment to practice your new skill a few times, then start showing off. You will be amazed at the reaction you get. (Whether or not you decide to reveal the secret is up to you!)

Welcome back. At this point, you probably have a few questions, such as:

"Can we use this method for multiplying three-digit numbers (or larger) by eleven?"

Absolutely. For instance, for the problem 314 × 11, the answer still begins with 3 and ends with 4. Since 3 + 1 = 4, and 1 + 4 = 5, the answer is 3454. But we'll save larger problems like this for later.

More practically, you are probably saying to yourself,

"Well, this is fine for multiplying by elevens, but what about larger numbers? How do I multiply numbers by twelve, or thirteen, or thirty-six?"

My answer to that is, *Patience!* That's what the rest of the book is all about. In Chapters 2, 3, 6, and 8, you will learn methods for multiplying together just about any two numbers. Better still, you don't have to memorize special rules for every number. Just a handful of techniques is all that it takes to multiply numbers in your head, quickly and easily.

SQUARING AND MORE

Here is another quick trick.

As you probably know, the square of a number is a number multiplied by itself. For example, the square of 7 is 7 × 7 = 49. Later, I will teach you a simple method that will enable you to easily calculate the square of any two-digit or three-digit (or higher) number. That method is especially simple when the number ends in 5, so let's do that trick now.

To square a two-digit number that ends in 5, you need to remember only two things.

1. The answer *begins* by multiplying the first digit by the next higher digit.
2. The answer *ends* in 25.

For example, to square the number 35, we simply multiply the first digit (3) by the next higher digit (4), then attach 25. Since 3 × 4 = 12, the answer is 1225. Therefore, 35 × 35 = 1225. Our steps can be illustrated this way:

$$
\begin{array}{r}
35 \\
\times\ 35 \\
\hline
3 \times 4 = 12 \\
5 \times 5 =\ \underline{\ 25} \\
\textbf{Answer: 1225}
\end{array}
$$

How about the square of 85? Since 8 × 9 = 72, we immediately get 85 × 85 = 7225.

$$
\begin{array}{r}
85 \\
\times\ 85 \\
\hline
8 \times 9 = 72 \\
5 \times 5 =\ \underline{\ 25} \\
\textbf{Answer: 7225}
\end{array}
$$

We can use a similar trick when multiplying two-digit numbers with the same first digit, and second digits that sum to 10. The answer begins the same way that it did before (the first digit multiplied by the next higher digit), followed by the product of the second digits. For example, let's try 83 × 87. (Both numbers begin with 8, and the last digits sum to 3 + 7 = 10.) Since 8 × 9 = 72, and 3 × 7 = 21, the answer is 7221.

$$
\begin{array}{r}
83 \\
\times\ 87 \\
\hline
8 \times 9 = 72 \\
3 \times 7 =\ \underline{\ 21} \\
\textbf{Answer: 7221}
\end{array}
$$

Similarly, $84 \times 86 = 7224$.

Now it's your turn. Try

$$26 \times 24$$

How does the answer begin? With $2 \times 3 = 6$. How does it end? With $6 \times 4 = 24$. Thus $26 \times 24 = 624$.

Remember that to use this method, the first digits have to be the same, and the last digits must sum to 10. Thus, we can use this method to instantly determine that

$$31 \times 39 = 1209$$
$$32 \times 38 = 1216$$
$$33 \times 37 = 1221$$
$$34 \times 36 = 1224$$
$$35 \times 35 = 1225$$

You may ask,

"What if the last digits do not sum to ten? Can we use this method to multiply twenty-two and twenty-three?"

Well, not yet. But in Chapter 8, I will show you an easy way to do problems like this using the close-together method. (For 22×23, you would do 20×25 plus 2×3, to get $500 + 6 = 506$, but I'm getting ahead of myself!) Not only will you learn how to use these methods, but you will understand why these methods work, too.

"Are there any tricks for doing mental addition and subtraction?"

Definitely, and that is what the next chapter is all about. If I were forced to summarize my method in three words, I would say, "Left to right." Here is a sneak preview.

Consider the subtraction problem

$$
\begin{array}{r}
1241 \\
-\ 587 \\
\end{array}
$$

Most people would not like to do this problem in their head (or even on paper!), but let's simplify it. Instead of subtracting 587, subtract 600. Since $1200 - 600 = 600$, we have that

$$
\begin{array}{r}
1241 \\
-\ 600 \\
\hline
641 \\
\end{array}
$$

But we have subtracted 13 too much. (We will explain how to quickly determine the 13 in Chapter 1.) Thus, our painful-looking subtraction problem becomes the easy addition problem

$$
\begin{array}{r}
641 \\
+\ 13 \\
\hline
654 \\
\end{array}
$$

which is not too hard to calculate in your head (especially from left to right). Thus, $1241 - 587 = 654$.

Using a little bit of mathematical magic, described in Chapter 9, you will be able to instantly compute the sum of the ten numbers below.

$$
\begin{array}{r}
9 \\
5 \\
14 \\
19 \\
33 \\
52 \\
85 \\
137 \\
222 \\
+\ 359 \\
\hline
935 \\
\end{array}
$$

Although I won't reveal the magical secret right now, here is a hint. The answer, 935, has appeared elsewhere in this chapter. More tricks for doing math on paper will be found in Chapter 6. Furthermore, you will be able to quickly give the quotient of the last two numbers:

$$359 \div 222 = 1.61 \text{ (first three digits)}$$

We will have much more to say about division (including decimals and fractions) in Chapter 4.

MORE PRACTICAL TIPS

Here's a quick tip for calculating tips. Suppose your bill at a restaurant came to $42, and you wanted to leave a 15% tip. First we calculate 10% of $42, which is $4.20. If we cut that number in half, we get $2.10, which is 5% of the bill. Adding these numbers together gives us $6.30, which is exactly 15% of the bill. We will discuss strategies for calculating sales tax, discounts, compound interest, and other practical items in Chapter 5, along with strategies that you can use for quick mental estimation when an exact answer is not required.

IMPROVE YOUR MEMORY

In Chapter 7, you will learn a useful technique for memorizing numbers. This will be handy in and out of the classroom. Using an easy-to-learn system for turning numbers into words, you will be able to quickly and easily memorize any numbers: dates, phone numbers, whatever you want.

Speaking of dates, how would you like to be able to figure out the day of the week of any date? You can use this to figure out birth dates, historical dates, future appointments, and so on. I will show you this in more detail later, but here is a simple way to figure out the day of January 1 for any year in the twenty-first century. First familiarize yourself with the following table.

Monday	Tuesday	Wednesday	Thursday	Friday	Saturday	Sunday
1	2	3	4	5	6	7 or 0

For instance, let's determine the day of the week of January 1, 2030. Take the last two digits of the year, and consider it to be your bill at a restaurant. (In this case, your bill would be $30.) Now add a 25% tip, but keep the change. (You can compute this by cutting the bill in half twice, and ignoring any change. Half of $30 is $15. Then half of $15 is $7.50. Keeping the change results in a $7 tip.) Hence your bill plus tip amounts to $37. To figure out the day of the week, subtract the biggest multiple of 7 (0, 7, 14, 21, 28, 35, 42, 49, . . .) from your total, and that will tell you the day of the week. In this case, $37 - 35 = 2$, and so January 1, 2030, will occur on 2's day, namely Tuesday:

$$
\begin{array}{rr}
\textbf{Bill:} & 30 \\
\textbf{Tip:} & +\ 7 \\
\hline
& 37 \\
\textbf{subtract 7s:} & -\ 35 \\
\hline
& 2 = \textbf{Tuesday}
\end{array}
$$

How about January 1, 2043:

$$
\begin{array}{rr}
\textbf{Bill:} & 43 \\
\textbf{Tip:} & +\ 10 \\
\hline
& 53 \\
\textbf{subtract 7s:} & -\ 49 \\
\hline
& 4 = \textbf{Thursday}
\end{array}
$$

Exception: If the year is a leap year, remove $1 from your tip, then proceed as before. For example, for January 1, 2032, a 25% tip of $32 would be $8. Removing one dollar gives a total of 32 + 7 = 39. Subtracting the largest multiple of 7 gives us 39 − 35 = 4. So January 1, 2032, will be on 4's day, namely Thursday. For more details that will allow you to compute the day of the week of any date in history, see Chapter 9. (In fact, it's perfectly okay to read that chapter first!)

I know what you are wondering now:

"Why didn't they teach this to us in school?"
I'm afraid that there are some questions that even I cannot answer. Are you ready to learn more magical math? Well, what are we waiting for? Let's go!

Chapter 1

A Little Give and Take:
Mental Addition and Subtraction

For as long as I can remember, I have always found it easier to add and subtract numbers from left to right instead of from right to left. By adding and subtracting numbers this way, I found that I could call out the answers to math problems in class well before my classmates put down their pencils. And I didn't even need a pencil!

In this chapter you will learn the left-to-right method of doing mental addition and subtraction for most numbers that you encounter on a daily basis. These mental skills are not only important for doing the tricks in this book but are also indispensable in school, at work, or any time you use numbers. Soon you will be able to retire your calculator and use the full capacity of your mind as you add and subtract two-digit, three-digit, and even four-digit numbers with lightning speed.

LEFT-TO-RIGHT ADDITION

Most of us are taught to do math on paper from right to left. And that's fine for doing math on paper. But if you want to do math *in your head* (even *faster* than you can on paper) there are many good reasons why it is better to work from left to right. After all, you read numbers from left to right, you pronounce numbers from left to right, and so it's just more natural to think about (and calculate) numbers from left to right. When you compute the answer from right to left (as you probably do on paper), you generate the answer backward. That's what makes it so hard to do math in your head. Also, if you want to estimate your answer, it's more important to know that your answer is "a little over 1200" than to know that your answer "ends in 8." Thus, by working from left to right, you begin with the most significant digits of your problem. If you are used to working from right to left on paper, it may seem unnatural to work with numbers from left to right. But with practice you will find that it is the most natural and efficient way to do mental calculations.

With the first set of problems—two-digit addition—the left-to-right method may not seem so advantageous. But be patient. If you stick with me, you will see that the only easy way to solve three-digit and larger addition problems, all subtraction problems, and most definitely all multiplication and division problems is from left to right. The sooner you get accustomed to computing this way, the better.

Two-Digit Addition

Our assumption in this chapter is that you know how to add and subtract one-digit numbers. We will begin with two-digit addition, something I suspect you can already do fairly well in your head. The following exercises are good practice, however, because the two-digit addition skills that you acquire here will be needed for larger addition problems, as well as virtually all multiplication problems in later chapters. It also illustrates a fundamental principle of mental arithmetic—namely, to *simplify* your problem by breaking it into smaller, more manageable parts. This is the key to virtually every method you will learn in this book. To paraphrase an old saying, there are three components to success—simplify, simplify, simplify.

The easiest two-digit addition problems are those that do not require you to *carry* any numbers, when the first digits sum to 9 or below and the last digits sum to 9 or below. For example:

$$47$$
$$\underline{+\ 32}\ (30 + 2)$$

To solve 47 + 32, first add 30, then add 2. After adding 30, you have the *simpler* problem 77 + 2, which equals 79. We illustrate this as follows:

$$47 + 32 \quad = \quad 77 + 2 \quad = \quad 79$$
(first add 30)   (then add 2)

The above diagram is simply a way of representing the mental processes involved in arriving at an answer using our method. While you need to be able to read and understand such diagrams as you work your way through this book, our method does not require you to write down *anything* yourself.

Now let's try a calculation that requires you to carry a number:

$$67$$
$$\underline{+\ 28}\ (20 + 8)$$

Adding from left to right, you can simplify the problem by adding 67 + 20 = 87; then 87 + 8 = 95.

$$67 + 28 \quad = \quad 87 + 8 \quad = \quad 95$$
(first add 20)   (then add 8)

Now try one on your own, mentally calculating from left to right, and then check below to see how we did it:

$$\begin{array}{r} 84 \\ + \ 57 \ (50 + 7) \\ \hline \end{array}$$

How was that? You added 84 + 50 = 134 and added 134 + 7 = 141.

$$84 + 57 \quad = \quad 134 + 7 \quad = \quad 141$$
(first add 50)   (then add 7)

If carrying numbers trips you up a bit, don't worry about it. This is probably the first time you have ever made a systematic attempt at mental calculation, and if you're like most people, it will take you time to get used to it. With practice, however, you will begin to see and hear these numbers in your mind, and carrying numbers when you add will come automatically. Try another problem for practice, again computing it in your mind first, then checking how we did it:

$$\begin{array}{r} 68 \\ + \ 45 \ (40 + 5) \\ \hline \end{array}$$

You should have added 68 + 40 = 108, and then 108 + 5 = 113, the final answer. Was that easier? If you would like to try your hand at more two-digit addition problems, check out the set of exercises below. (The answers and computations are at the end of the book.)

EXERCISE: TWO-DIGIT ADDITION

1.	2.	3.	4.	5.
23 + 16	64 + 43	95 + 32	34 + 26	89 + 78

6.	7.	8.	9.	10.
73 + 58	47 + 36	19 + 17	55 + 49	39 + 38

Three-Digit Addition

The strategy for adding three-digit numbers is the same as for adding two-digit numbers: you add from left to right. After each step, you arrive at a new (and *simpler*) addition problem. Let's try the following:

$$538$$
$$+ 327 \ (300 + 20 + 7)$$

Starting with 538, we add 300, then add 20, then add 7. After adding 300 (538 + 300 = 838), the problem becomes 838 + 27. After adding 20 (838 + 20 = 858), the problem simplifies to 858 + 7 = 865. This thought process can be diagrammed as follows:

$$538 + 327 \quad \underset{+ \ 300}{=} \quad 838 + 27 \quad \underset{+ \ 20}{=} \quad 858 + 7 \quad \underset{+ \ 7}{=} \quad 865$$

All mental addition problems can be done by this method. The goal is to keep simplifying the problem until you are just adding a one-digit number. Notice that 538 + 327 requires you to hold on to six digits in your head, whereas 838 + 27 and

858 + 7 require only five and four digits, respectively. *As you simplify the problem, the problem gets easier!*

Try the following addition problem in your mind before looking to see how we did it:

$$623$$
$$+ 159 \ (100 + 50 + 9)$$

Did you reduce and simplify the problem by adding left to right? After adding the hundreds (623 + 100 = 723), you were left with 723 + 59. Next you should have added the tens (723 + 50 = 773), simplifying the problem to 773 + 9, which you then summed to get 782. Diagrammed, the problem looks like this:

$$623 + 159 \quad = \quad 723 + 59 \quad = \quad 773 + 9 \quad = \quad 782$$
$$\quad\quad + 100 \quad\quad\quad\quad + 50 \quad\quad\quad\quad + 9$$

When I do these problems mentally, I do not try to *see* the numbers in my mind—I try to *hear* them. I hear the problem 623 + 159 as *six hundred twenty-three* plus *one hundred fifty-nine;* by emphasizing the word *hundred* to myself, I know where to begin adding. Six plus one equals seven, so my next problem is *seven hundred and twenty-three* plus *fifty-nine,* and so on. When first doing these problems, practice them out loud. Reinforcing yourself verbally will help you learn the mental method much more quickly.

Three-digit addition problems really do not get much harder than the following:

$$858$$
$$+ 634$$

Now look to see how we did it:

$$858 + 634 \underset{+\,600}{=} 1458 + 34 \underset{+\,30}{=} 1488 + 4 \underset{+\,4}{=} 1492$$

At each step I hear (not see) a "new" addition problem. In my mind the problem sounds like this:

858 plus 634 is 1458 plus 34 is 1488 plus 4 is 1492.

Your mind-talk may not sound exactly like mine (indeed, you might "see" the numbers instead of "hear" them), but whatever it is you say or visualize to yourself, the point is to reinforce the numbers along the way so that you don't forget where you are and have to start the addition problem over again.

Let's try another one for practice:

$$
\begin{array}{r}
759 \\
+\,496 \;(400 + 90 + 6) \\
\hline
\end{array}
$$

Do it in your mind first, then check our computation below:

$$759 + 496 \underset{+\,400}{=} 1159 + 96 \underset{+\,90}{=} 1249 + 6 \underset{+\,6}{=} 1255$$

This addition problem is a little more difficult than the last one since it requires you to carry numbers in all three steps. However, with this particular problem you have the option of using an alternative method. I am sure you will agree that it is a lot easier to add 500 to 759 than it is to add 496, so try adding 500 and then subtracting the difference:

$$759$$
$$+ \ 496 \ (500 - 4)$$

$$759 + 496 \ = \ 1259 - 4 \ = \ 1255$$
$$\text{(first add 500)} \qquad \text{(then subtract 4)}$$

So far, you have consistently broken up the second number in any problem to add to the first. It really does not matter which number you choose to break up, but it is good to be consistent. That way, your mind will never have to waste time deciding which way to go. If the second number happens to be a lot simpler than the first, I sometimes switch them around, as in the following example:

$$207$$
$$+ \ 528$$

$$207 + 528 \ = \ 528 + 207 \ = \ 728 + 7 \ = \ 735$$
$$\text{(switch)} \qquad + \ 200 \qquad \quad + \ 7$$

Let's finish up by adding three-digit to four-digit numbers. Since most human memory can hold only about seven or eight digits at a time, this is about as large a problem as you can handle without resorting to artificial memory devices, like fingers, calculators, or the mnemonics taught in Chapter 7. In many addition problems that arise in practice, especially within multiplication problems, one or both of the numbers will end in 0, so we shall emphasize those types of problems. We begin with an easy one:

$$2700$$
$$+ \ 567$$

Since 27 *hundred* + 5 *hundred* is 32 *hundred,* we simply attach the 67 to get 32 *hundred* and 67, or 3267. The process is the same for the following problems:

$$
\begin{array}{r}
3240 \\
+\ \ 18 \\
\hline
\end{array}
\qquad
\begin{array}{r}
3240 \\
+\ \ 72 \\
\hline
\end{array}
$$

Because 40 + 18 = 58, the first answer is 3258. For the second problem, since 40 + 72 exceeds 100, you know the answer will be 33 hundred and *something*. Because 40 + 72 = 112, the answer is 3312.

These problems are easy because the (nonzero) digits overlap in only one place, and hence can be solved in a single step. Where digits overlap in two places, you require two steps. For instance:

$$
\begin{array}{r}
4560 \\
+\ \ 171\ (100 + 71) \\
\hline
\end{array}
$$

This problem requires two steps, as diagrammed the following way:

$$
\underset{+\,100}{4560 + 171} \quad = \quad \underset{+\,71}{4660 + 71} \quad = \quad 4731
$$

Practice the following three-digit addition exercises, and then add some (pun intended!) of your own if you like until you are comfortable doing them mentally without having to look down at the page. (Answers can be found in the back of the book.)

Carl Friedrich Gauss: Mathematical Prodigy

A prodigy is a highly talented child, usually called precocious or gifted, and almost always ahead of his peers. The German mathematician Carl Friedrich Gauss (1777–1855) was one such child. He often boasted that he could calculate before he could speak. By the ripe old age of three, before he had been taught any arithmetic, he corrected his father's payroll by declaring "the reckoning is wrong." A further check of the numbers proved young Carl correct.

As a ten-year-old student, Gauss was presented the following mathematical problem: What is the sum of numbers from 1 to 100? While his fellow students were frantically calculating with paper and pencil, Gauss immediately envisioned that if he spread out the numbers 1 through 50 from left to right, and the numbers 51 to 100 from right to left directly below the 1–50 numbers, each combination would add up to 101 (1 + 100, 2 + 99, 3 + 98, . . .). Since there were fifty sums, the answer would be 101 × 50 = 5050. To the astonishment of everyone, including the teacher, young Carl got the answer not only ahead of everyone else, but computed it entirely in his mind. He wrote out the answer on his slate, and flung it on the teacher's desk with a defiant "There it lies." The teacher was so impressed that he invested his own money to purchase the best available textbook on arithmetic and gave it to Gauss, stating, "He is beyond me, I can teach him nothing more."

Indeed, Gauss became the mathematics teacher of others, and eventually went on to become one of the greatest mathematicians in history, his theories still used today in the service of science. Gauss's desire to better understand Nature through the language of mathematics was summed up in his motto, taken from Shakespeare's *King Lear* (substituting "laws" for "law"): "Thou, nature, art my goddess; to thy laws/My services are bound."

EXERCISE: THREE-DIGIT ADDITION

1.	2.	3.	4.	5.
242 + 137	312 + 256	635 + 814	457 + 241	912 + 475

6.	852	7.	457	8.	878	9.	276	10.	877
	+ 378		+ 269		+ 797		+ 689		+ 539

11.	5400	12.	1800	13.	6120	14.	7830	15.	4240
	+ 252		+ 855		+ 136		+ 348		+ 371

LEFT-TO-RIGHT SUBTRACTION

For most of us, it is easier to add than to subtract. But if you continue to compute from left to right and to break down problems into simpler components, subtraction can become almost as easy as addition.

Two-Digit Subtraction

When subtracting two-digit numbers, your goal is to simplify the problem so that you are reduced to subtracting (or adding) a one-digit number. Let's begin with a very simple subtraction problem:

$$
\begin{array}{r}
86 \\
- \ 25 \ (20 + 5) \\
\hline
\end{array}
$$

After each step, you arrive at a new and easier subtraction problem. Here, we first subtract 20 (86 − 20 = 66) then we subtract 5 to reach the simpler subtraction problem 66 − 5 for your final answer of 61. The problem can be diagrammed this way:

$$86 - 25 \ = \ 66 - 5 \ = \ 61$$
$$\text{(first subtract 20)} \quad \text{(then subtract 5)}$$

Of course, subtraction problems are considerably easier when there is no borrowing (which occurs when a larger digit is being subtracted from a smaller one). But the good news is that "hard" subtraction problems can usually be turned into "easy" addition problems. For example:

$$86$$
$$\underline{-\ 29}\ (20 + 9)\ \text{or}\ (30 - 1)$$

There are two different ways to solve this problem mentally:

1. First subtract 20, then subtract 9:

$$86 - 29\ \underset{\text{(first subtract 20)}}{=}\ 66 - 9\ \underset{\text{(then subtract 9)}}{=}\ 57$$

But for this problem, I would prefer the following strategy:

2. First subtract 30, then add back 1:

$$86 - 29\ \underset{\text{(first subtract 30)}}{=}\ 56 + 1\ \underset{\text{(then add 1)}}{=}\ 57$$

Here is the rule for deciding which method to use: If a two-digit subtraction problem would require borrowing, then round the second number up (to a multiple of ten). Subtract the rounded number, then add back the difference.

For example, the problem $54 - 28$ would require borrowing (since 8 is greater than 4), so round 28 up to 30, compute $54 - 30 = 24$, then add back 2 to get 26 as your final answer:

$$54$$
$$\underline{-\ 28}\ (30 - 2)$$

$$54 - 28\ \underset{-\ 30}{=}\ 24 + 2\ \underset{+\ 2}{=}\ 26$$

Now try your hand (or head) at 81 − 37. Since 7 is greater than 1, we round 37 up to 40, subtract it from 81 (81 − 40 = 41), then add back the difference of 3 to arrive at the final answer:

$$81 - 37 \underset{-40}{=} 41 + 3 \underset{+3}{=} 44$$

With just a little bit of practice, you will become comfortable working subtraction problems both ways. Just use the rule above to decide which method will work best.

EXERCISE: TWO-DIGIT SUBTRACTION

1.		2.		3.		4.		5.	
	38		**84**		**92**		**67**		**79**
	− 23		**− 59**		**− 34**		**− 48**		**− 29**

6.		7.		8.		9.		10.	
	63		**51**		**89**		**125**		**148**
	− 46		**− 27**		**− 48**		**− 79**		**− 86**

Three-Digit Subtraction

Now let's try a three-digit subtraction problem:

$$\begin{array}{r} 958 \\ - 417 \ (400 + 10 + 7) \end{array}$$

This particular problem does not require you to borrow any numbers (since every digit of the second number is less than the digit above it), so you should not find it too hard. Simply subtract one digit at a time, simplifying as you go.

$$958 - 417 \underset{-400}{=} 558 - 17 \underset{-10}{=} 548 - 7 \underset{-7}{=} 541$$

Now let's look at a three-digit subtraction problem that requires you to borrow a number:

$$747$$
$$\underline{- 598} \text{ (600 − 2)}$$

At first glance this probably looks like a pretty tough problem, but if you first subtract 747 − 600 = 147, then add back 2, you reach your final answer of 147 + 2 = 149.

$$747 - 598 \underset{-600}{=} 147 + 2 \underset{+2}{=} 149$$

Now try one yourself:

$$853$$
$$\underline{- 692}$$

Did you first subtract 700 from 853? If so, did you get 853 − 700 = 153? Since you subtracted by 8 too much, did you add back 8 to reach 161, the final answer?

$$853 - 692 \underset{-700}{=} 153 + 8 \underset{+8}{=} 161$$

Now, I admit we have been making life easier for you by subtracting numbers that were close to a multiple of 100. (Did you notice?) But what about other problems, like:

$$725$$
$$\underline{- 468} \text{ (400 + 60 + 8) or (500 − ??)}$$

If you subtract one digit at a time, simplifying as you go, your sequence will look like this:

725 − 468   =   325 − 68   =   265 − 8   =   257
(first subtract 400)   (then subtract 60)   (then subtract 8)

What happens if you round up to 500?

725 − 468   =   225 + ??   =   ??
(first subtract 500)   (then add ??)

Subtracting 500 is easy: $725 − 500 = 225$. But you have subtracted too much. The trick is to figure out exactly how much too much.

At first glance, the answer is far from obvious. To find it, you need to know how far 468 is from 500. The answer can be found by using "complements," a nifty technique that will make many three-digit subtraction problems a lot easier to do.

Using Complements (You're Welcome!)

Quick, how far from 100 are each of these numbers?

57   68   49   21   79

Here are the answers:

57	68	49	21	79
+ 43	+ 32	+ 51	+ 79	+ 21
100	100	100	100	100

Notice that for each pair of numbers that add to 100, the first digits (on the left) add to 9 and the last digits (on the right) add to 10. We say that 43 is the complement of 57, 32 is the complement of 68, and so on.

Now you find the complements of these two-digit numbers:

37   59   93   44   08

To find the complement of 37, first figure out what you need to add to 3 in order to get 9. (The answer is 6.) Then figure out what you need to add to 7 to get 10. (The answer is 3.) Hence, 63 is the complement of 37.

The other complements are 41, 7, 56, 92. Notice that, like everything else you do as a mathemagician, the complements are determined from left to right. As we have seen, the first digits add to 9, and the second digits add to 10. (An exception occurs in numbers ending in 0—e.g., 30 + 70 = 100—but those complements are simple!)

What do complements have to do with mental subtraction? Well, they allow you to convert difficult subtraction problems into straightforward addition problems. Let's consider the last subtraction problem that gave us some trouble:

725
− 468 (500 − 32)

To begin, you subtracted 500 instead of 468 to arrive at 225 (725 − 500 = 225). But then, having subtracted too much, you needed to figure out how much to add back. Using complements gives you the answer in a flash. How far is 468 from 500? The same distance as 68 is from 100. If you find the complement of 68 the way we have shown you, you will arrive at 32. Add 32 to 225, and you will arrive at 257, your final answer.

$$725 - 468 \quad = \quad 225 + 32 \quad = \quad 257$$
(first subtract 500)   (then add 32)

Try another three-digit subtraction problem:

$$821$$
$$\underline{- 259} \ (300 - 41)$$

To compute this mentally, subtract 300 from 821 to arrive at 521, then add back the complement of 59, which is 41, to arrive at 562, our final answer. The procedure looks like this:

$$821 - 259 \quad = \quad 521 + 41 \quad = \quad 562$$
$$- 300 \qquad\qquad\quad + 41$$

Here is another problem for you to try:

$$645$$
$$\underline{- 372} \ (400 - 28)$$

Check your answer and the procedure for solving the problem below:

$$645 - 372 \quad = \quad 245 + 28 \quad = \quad 265 + 8 \quad = \quad 273$$
$$- 400 \qquad\qquad + 20 \qquad\qquad + 8$$

Subtracting a three-digit number from a four-digit number is not much harder, as the next example illustrates:

$$1246$$
$$\underline{- \ 579} \ (600 - 21)$$

By rounding up, you subtract 600 from 1246, leaving 646, then add back the complement of 79, which is 21. Your final answer is 646 + 21 = 667.

$$1246 - 579 \underset{-600}{=} 646 + 21 \underset{+21}{=} 667$$

Try the three-digit subtraction exercises below, and then create more of your own for additional (or should that be subtractional?) practice.

EXERCISE: THREE-DIGIT SUBTRACTION

1.	583	2.	936	3.	587	4.	763	5.	204
	− 271		− 725		− 298		− 486		− 185
6.	793	7.	219	8.	978	9.	455	10.	772
	− 402		− 176		− 784		− 319		− 596
11.	873	12.	564	13.	1428	14.	2345	15.	1776
	− 357		− 228		− 571		− 678		− 987

Products of a Misspent Youth: Basic Multiplication

I probably spent too much time of my childhood thinking about faster and faster ways to perform mental multiplication; I was diagnosed as hyperactive and my parents were told that I had a short attention span and probably would not be successful in school. (Fortunately, my parents ignored that advice. I was also lucky to have some incredibly patient teachers in my first few years of school.) It might have been my short attention span that motivated me to develop quick ways to do arithmetic. I don't think I had the patience to carry out math problems with pencil and paper. Once you have mastered the techniques described in this chapter, you won't want to rely on pencil and paper again, either.

In this chapter you will learn how to multiply in your head one-digit numbers by two-digit numbers and three-digit numbers. You will also learn a phenomenally fast way to square two-digit numbers. Even friends with calculators won't be able to keep up with you. Believe me, virtually everyone will be dumbfounded by the fact that such problems can not only be done mentally, but can be computed so quickly. I sometimes wonder whether we were not cheated in school; these methods are so simple once you learn them.

There is one small prerequisite for mastering the skills in this chapter—you need to know the multiplication tables through ten. In fact, to really make headway, you need to know your multiplication tables backward and forward. For those of you who need to shake the cobwebs loose, consult the multiplication chart below. Once you've got your tables down, you are ready to begin.

Multiplication Table of Numbers 1–10

×	1	2	3	4	5	6	7	8	9	10
1	1	2	3	4	5	6	7	8	9	10
2	2	4	6	8	10	12	14	16	18	20
3	3	6	9	12	15	18	21	24	27	30
4	4	8	12	16	20	24	28	32	36	40
5	5	10	15	20	25	30	35	40	45	50
6	6	12	18	24	30	36	42	48	54	60
7	7	14	21	28	35	42	49	56	63	70
8	8	16	24	32	40	48	56	64	72	80
9	9	18	27	36	45	54	63	72	81	90
10	10	20	30	40	50	60	70	80	90	100

2-BY-1 MULTIPLICATION PROBLEMS

If you worked your way through Chapter 1, you got into the habit of adding and subtracting from left to right. You will do virtually all the calculations in this chapter from left to right as well. This is undoubtedly the opposite of what you learned in school.

But you'll soon see how much easier it is to think from left to right than from right to left. (For one thing, you can start to say your answer aloud before you have finished the calculation. That way you seem to be calculating even faster than you are!)

Let's tackle our first problem:

$$
\begin{array}{r}
42 \\
\times\ 7 \\
\hline
\end{array}
$$

First, multiply $40 \times 7 = 280$. (Note that 40×7 is just like 4×7, with a friendly zero attached.) Next, multiply $2 \times 7 = 14$. Then add 280 plus 14 (left to right, of course) to arrive at 294, the correct answer. We illustrate this procedure below.

$$
\begin{array}{r}
42\ (40 + 2) \\
\times\ \ 7 \\
\hline
40 \times 7 = \quad 280 \\
2 \times 7 = +\ 14 \\
\hline
294
\end{array}
$$

We have omitted diagramming the mental addition of 280 + 14, since you learned how to do that computation in the last chapter. At first you will need to look down at the problem while doing the calculation. With practice you will be able to forgo this step and compute the whole thing in your mind.

Let's try another example:

$$
\begin{array}{r}
48\ (40 + 8) \\
\times\ 4 \\
\hline
\end{array}
$$

Your first step is to break down the problem into small multiplication tasks that you can perform mentally with ease. Since

48 = 40 + 8, multiply 40 × 4 = 160, then add 8 × 4 = 32. The answer is 192. (Note: If you are wondering *why* this process works, see the Why These Tricks Work section at the end of the chapter.)

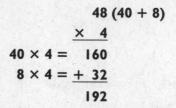

$$
\begin{array}{r}
48\ (40 + 8) \\
\times\ \ 4 \\
\hline
40 \times 4 = \quad 160 \\
8 \times 4 = +\ 32 \\
\hline
192
\end{array}
$$

Here are two more mental multiplication problems that you should be able to solve fairly quickly. First calculate 62 × 3. Then do 71 × 9. Try doing them in your head before looking at how we did it.

$$
\begin{array}{r}
62\ (60 + 2) \\
\times\ \ 3 \\
\hline
60 \times 3 = \quad 180 \\
2 \times 3 = +\ 6 \\
\hline
186
\end{array}
\qquad
\begin{array}{r}
71\ (70 + 1) \\
\times\ \ 9 \\
\hline
70 \times 9 = \quad 630 \\
1 \times 9 = +\ 9 \\
\hline
639
\end{array}
$$

These two examples are especially simple because the numbers being added essentially do not overlap at all. When doing 180 + 6, you can practically *hear* the answer: *One hundred eighty . . . six!* Another especially easy type of mental multiplication problem involves numbers that begin with five. When the five is multiplied by an even digit, the first product will be a multiple of 100, which makes the resulting addition problem a snap.

$$58 \ (50 + 8)$$
$$\underline{\times \quad 4}$$

$$50 \times 4 = \quad 200$$
$$8 \times 4 = \underline{+ \ 32}$$
$$232$$

Try your hand at the following problem:

$$87 \ (80 + 7)$$
$$\underline{\times \quad 5}$$

$$80 \times 5 = \quad 400$$
$$7 \times 5 = \underline{+ \ 35}$$
$$435$$

Notice how much easier this problem is to do from left to right. It takes far less time to calculate "400 plus 35" mentally than it does to apply the pencil-and-paper method of "putting down the 5 and carrying the 3."

The following two problems are a little harder.

$38 \ (30 + 8)$	$67 \ (60 + 7)$
$\underline{\times \quad 9}$	$\underline{\times \quad 8}$
$30 \times 9 = \quad 270$	$60 \times 8 = \quad 480$
$8 \times 9 = \underline{+ \ 72}$	$7 \times 8 = \underline{+ \ 56}$
342	536

As usual, we break these problems down into easier problems. For the one on the left, multiply 30×9 plus 8×9, giving you $270 + 72$. The addition problem is slightly harder because it involves carrying a number. Here $270 + 70 + 2 = 340 + 2 = 342$.

With practice, you will become more adept at juggling

problems like these in your head, and those that require you to carry numbers will be almost as easy as those that don't.

Rounding Up

You saw in the last chapter how useful rounding up can be when it comes to subtraction. The same goes for multiplication, especially when you are multiplying numbers that end in eight or nine.

Let's take the problem of 69 × 6, illustrated below. On the left we have calculated it the usual way, by adding 360 + 54. On the right, however, we have rounded 69 up to 70, and subtracted 420 − 6, which you might find easier to do.

$$
\begin{array}{rr}
69\,(60+9) & \quad or \\
\times\ 6 & \\
60\times6=\ \ \ 360 & \\
9\times6=+\ 54 & \\
\hline
414 &
\end{array}
\qquad
\begin{array}{rr}
69\,(70-1) \\
\times\ 6 \\
70\times6=\ \ \ 420 \\
-1\times6=-\ \ 6 \\
\hline
414
\end{array}
$$

The following example also shows how much easier rounding up can be:

$$
\begin{array}{rr}
78\,(70+8) & \quad or \\
\times\ 9 & \\
70\times9=\ \ \ 630 & \\
8\times9=+\ 72 & \\
\hline
702 &
\end{array}
\qquad
\begin{array}{rr}
78\,(80-2) \\
\times\ 9 \\
80\times9=\ \ \ 720 \\
-2\times9=-\ 18 \\
\hline
702
\end{array}
$$

The subtraction method works especially well for numbers that are just one or two digits away from a multiple of 10. It does not work so well when you need to round up more than two digits because the subtraction portion of the problem gets difficult. As it is, you may prefer to stick with the addition method. Personally, for problems of this size, I use only the addition method because in the time spent deciding which method to use, I could have already done the calculation!

So that you can perfect your technique, I strongly recommend practicing more 2-by-1 multiplication problems. Below are twenty problems for you to tackle. I have supplied you with the answers in the back, including a breakdown of each component of the multiplication. If, after you've worked out these problems, you would like to practice more, make up your own. Calculate mentally, then check your answer with a calculator. Once you feel confident that you can perform these problems rapidly in your head, you are ready to move to the next level of mental calculation.

EXERCISE: 2-BY-1 MULTIPLICATION

1. 82 × 9	2. 43 × 7	3. 67 × 5	4. 71 × 3	5. 93 × 8
6. 49 × 9	7. 28 × 4	8. 53 × 5	9. 84 × 5	10. 58 × 6
11. 97 × 4	12. 78 × 2	13. 96 × 9	14. 75 × 4	15. 57 × 7
16. 37 × 6	17. 46 × 2	18. 76 × 8	19. 29 × 3	20. 64 × 8

3-BY-1 MULTIPLICATION PROBLEMS

Now that you know how to do 2-by-1 multiplication problems in your head, you will find that multiplying three digits by a single digit is not much more difficult. You can get started with the following 3-by-1 problem (which is really just a 2-by-1 problem in disguise):

$$
\begin{array}{r}
320\ (300 + 20) \\
\times\ \ \ \ 7 \\
\hline
300 \times 7 =\ \ \ 2100 \\
20 \times 7 =\ +\ 140 \\
\hline
2240
\end{array}
$$

Was that easy for you? (If this problem gave you trouble, you might want to review the addition material in Chapter 1.) Let's try another 3-by-1 problem similar to the one you just did, except we have replaced the 0 with a 6 so you have another step to perform:

$$
\begin{array}{r}
326\ (300 + 20 + 6) \\
\times\ \ \ \ 7 \\
\hline
300 \times 7 =\ \ \ 2100 \\
20 \times 7 =\ +\ 140 \\
\hline
2240 \\
6 \times 7 =\ +\ \ \ \ 42 \\
\hline
2282
\end{array}
$$

In this case, you simply add the product of 6 × 7, which you already know to be 42, to the first sum of 2240. Since you do not need to carry any numbers, it is easy to add 42 to 2240 to arrive at the total of 2282.

In solving this and other 3-by-1 multiplication problems, the difficult part may be holding in memory the first sum (in this case, 2240) while doing the next multiplication problem (in this case, 6 × 7). There is no magic secret to remembering that first number, but with practice I guarantee you will improve your concentration, and holding on to numbers while performing other functions will get easier.

Let's try another problem:

$$
\begin{array}{r}
647 \ (600 + 40 + 7) \\
\times \quad 4 \\
\end{array}
$$

$$
\begin{array}{r}
600 \times 4 = \quad 2400 \\
40 \times 4 = + \ 160 \\
\hline
2560 \\
7 \times 4 = + \quad 28 \\
\hline
2588 \\
\end{array}
$$

Even if the numbers are large, the process is just as simple. For example:

$$
\begin{array}{r}
987 \ (900 + 80 + 7) \\
\times \quad 9 \\
\end{array}
$$

$$
\begin{array}{r}
900 \times 9 = \quad 8100 \\
80 \times 9 = + \ 720 \\
\hline
8820 \\
7 \times 9 = + \quad 63 \\
\hline
8883 \\
\end{array}
$$

When first solving these problems, you may have to glance down at the page as you go along to remind yourself what the original problem is. This is okay at first. But try to break the habit so that eventually you are holding the problem entirely in memory.

In the last section on 2-by-1 multiplication problems, we saw that problems involving numbers that begin with five are sometimes especially easy to solve. The same is true for 3-by-1 problems:

$$
\begin{array}{r}
563\ (500 + 60 + 3) \\
\times\quad 6 \\
\hline
500 \times 6 = \quad 3000 \\
60 \times 6 = \quad 360 \\
3 \times 6 = +\ \ 18 \\
\hline
3378
\end{array}
$$

Notice that whenever the first product is a multiple of 1000, the resulting addition problem is no problem at all. This is because you do not have to carry any numbers and the thousands digit does not change. If you were solving the problem above in front of someone else, you would be able to say your first product—"three thousand . . ."—out loud with complete confidence that a carried number would not change it to 4000. (As an added bonus, by quickly saying the first digit, it gives the illusion that you computed the entire answer immediately!) Even if you are practicing alone, saying your first product out loud frees up some memory space while you work on the remaining 2-by-1 problem, which you can say out loud as well—in this case, ". . . three hundred seventy-eight."

Try the same approach in solving the next problem, where the multiplier is a 5:

$$
\begin{array}{r}
663\ (600 + 60 + 3) \\
\times\quad 5 \\
\hline
600 \times 5 = \quad 3000 \\
60 \times 5 = \quad 300 \\
3 \times 5 = +\ \ 15 \\
\hline
3315
\end{array}
$$

Because the first two digits of the three-digit number are even, you can say the answer as you calculate it without having to add anything! Don't you wish all multiplication problems were this easy?

Let's escalate the challenge by trying a couple of problems that require some carrying.

$$
\begin{array}{r}
184\ (100 + 80 + 4) \\
\times\quad 7 \\
\end{array}
$$

$100 \times 7 =$	700
$80 \times 7 =$	+ 560
	1260
$4 \times 7 =$	+ 28
	1288

$$
\begin{array}{r}
684\ (600 + 80 + 4) \\
\times\quad 9 \\
\end{array}
$$

$600 \times 9 =$	5400
$80 \times 9 =$	+ 720
	6120
$4 \times 9 =$	+ 36
	6156

In the next two problems you need to carry a number at the end of the problem instead of at the beginning:

$$
\begin{array}{r}
648\ (600 + 40 + 8) \\
\times\quad 9 \\
\end{array}
$$

$600 \times 9 =$	5400
$40 \times 9 =$	+ 360
	5760
$8 \times 9 =$	+ 72
	5832

$$
\begin{array}{r}
376\ (300 + 70 + 6) \\
\times \quad\ 4 \\
\hline
\end{array}
$$

$$
\begin{array}{rr}
300 \times 4 = & 1200 \\
70 \times 4 = & +\ 280 \\
\hline
& 1480 \\
6 \times 4 = & +\quad 24 \\
\hline
& 1504 \\
\end{array}
$$

The first part of each of these problems is easy enough to compute mentally. The difficult part comes in holding the preliminary answer in your head while computing the final answer. In the case of the first problem, it is easy to add 5400 + 360 = 5760, but you may have to repeat 5760 to yourself several times while you multiply 8 × 9 = 72. Then add 5760 + 72. Sometimes at this stage I will start to say my answer aloud before finishing. Because I know I will have to carry when I add 60 + 72, I know that 5700 will become 5800, so I say "fifty-eight hundred and . . ." Then I pause to compute 60 + 72 = 132. Because I have already carried, I say only the last two digits, ". . . thirty-two!" And there is the answer: 5832.

The next two problems require you to carry two numbers each, so they may take you longer than those you have already done. But with practice you will get faster:

$$
\begin{array}{r}
489\ (400 + 80 + 9) \\
\times \quad\ 7 \\
\hline
\end{array}
$$

$$
\begin{array}{rr}
400 \times 7 = & 2800 \\
80 \times 7 = & +\ 560 \\
\hline
& 3360 \\
9 \times 7 = & +\quad 63 \\
\hline
& 3423 \\
\end{array}
$$

$$224 \ (200 + 20 + 4)$$
$$\times \ \ \ \ 9$$
$$200 \times 9 = \ \ \ 1800$$
$$20 \times 9 = + \ 180$$
$$1980$$
$$4 \times 9 = + \ \ \ 36$$
$$2016$$

When you are first tackling these problems, repeat the answers to each part out loud as you compute the rest. In the first problem, for example, start by saying, "Twenty-eight hundred plus five hundred sixty" a couple of times out loud to reinforce the two numbers in memory while you add them together. Repeat the answer—"thirty-three hundred sixty"— several times while you multiply $9 \times 7 = 63$. Then repeat "thirty-three hundred sixty plus sixty-three" aloud until you compute the final answer of 3423. If you are thinking fast enough to recognize that adding $60 + 63$ will require you to carry a 1, you can begin to give the final answer a split second before you know it—"thirty-four hundred and . . . twenty-three!"

Let's end this section on 3-by-1 multiplication problems with some special problems you can do in a flash because they require one addition step instead of two:

$$511 \ (500 + 11)$$
$$\times \ \ \ \ 7$$
$$500 \times 7 = \ \ \ 3500$$
$$11 \times 7 = + \ \ \ 77$$
$$3577$$

$$
\begin{array}{r}
925\ (900 + 25) \\
\times\quad 8 \\
\hline
\end{array}
$$

$$
\begin{array}{rr}
900 \times 8 = & 7200 \\
25 \times 8 = & +\ 200 \\
\hline
& 7400
\end{array}
$$

$$
\begin{array}{r}
825\ (800 + 25) \\
\times\quad 3 \\
\hline
\end{array}
$$

$$
\begin{array}{rr}
800 \times 3 = & 2400 \\
25 \times 3 = & +\quad 75 \\
\hline
& 2475
\end{array}
$$

In general, if the product of the last two digits of the first number and the multiplier is known to you without having to calculate it (for instance, you may know that $25 \times 8 = 200$ automatically since 8 quarters equals $2.00), you will get to the final answer much more quickly. For instance, if you know without calculating that $75 \times 4 = 300$, then it is easy to compute 975×4:

$$
\begin{array}{r}
975\ (900 + 75) \\
\times\quad 4 \\
\hline
\end{array}
$$

$$
\begin{array}{rr}
900 \times 4 = & 3600 \\
75 \times 4 = & +\ 300 \\
\hline
& 3900
\end{array}
$$

To reinforce what you have just learned, solve the following 3-by-1 multiplication problems in your head; then check your computations and answers with ours (in the back of the book). I can assure you from experience that doing mental calculations is just like riding a bicycle or typing. It might seem impossible at first, but once you've mastered it, you will never forget how to do it.

EXERCISE: 3-BY-1 MULTIPLICATION

1. 431
    × 6

2. 637
    × 5

3. 862
    × 4

4. 957
    × 6

5. 927
    × 7

6. 728
    × 2

7. 328
    × 6

8. 529
    × 9

9. 807
    × 9

10. 587
    × 4

11. 184
    × 7

12. 214
    × 8

13. 757
    × 8

14. 259
    × 7

15. 297
    × 8

16. 751
    × 9

17. 457
    × 7

18. 339
    × 8

19. 134
    × 8

20. 611
    × 3

21. 578
    × 9

22. 247
    × 5

23. 188
    × 6

24. 968
    × 6

25. 499
    × 9

26. 670
    × 4

27. 429
    × 3

28. 862
    × 5

29. 285
    × 6

30. 488
    × 9

31. 693
    × 6

32. 722
    × 9

33. 457
    × 9

34. 767
    × 3

35. 312
    × 9

36. 691
    × 3

BE THERE OR B²: SQUARING TWO-DIGIT NUMBERS

Squaring numbers in your head (multiplying a number by itself) is one of the easiest yet most impressive feats of mental calculation you can do. I can still recall where I was when I discovered how to do it. I was thirteen, sitting on a bus on the way to visit my father at work in downtown Cleveland. It was a trip I made often, so my mind began to wander. I'm not sure why, but I began thinking about the numbers that add up to 20, and I wondered, how large could the product of two such numbers get?

I started in the middle with 10×10 (or 10^2), the product of which is 100. Next, I multiplied $9 \times 11 = 99$, $8 \times 12 = 96$, $7 \times 13 = 91$, $6 \times 14 = 84$, $5 \times 15 = 75$, $4 \times 16 = 64$, and so on. I noticed that the products were getting smaller, and their difference from 100 was 1, 4, 9, 16, 25, 36, . . . —or $1^2, 2^2, 3^2, 4^2, 5^2, 6^2, \ldots$ (see table below).

Numbers that add to 20		Distance from 10	Their product	Product's difference from 100
10	10	0	100	0
9	11	1	99	1
8	12	2	96	4
7	13	3	91	9
6	14	4	84	16
5	15	5	75	25
4	16	6	64	36
3	17	7	51	49
2	18	8	36	64
1	19	9	19	81

I found this pattern astonishing. Next I tried numbers that add to 26 and got similar results. First I worked out $13^2 = 169$, then computed $12 \times 14 = 168$, $11 \times 15 = 165$, $10 \times 16 = 160$,

$9 \times 17 = 153$, and so on. Just as before, the distances these products were from 169 was 1^2, 2^2, 3^2, 4^2, and so on (see table below).

There is actually a simple algebraic explanation for this phenomenon (see Why These Tricks Work, page 50). At the time, I didn't know my algebra well enough to prove that this pattern would always occur, but I experimented with enough examples to become convinced of it.

Then I realized that this pattern could help me square numbers more easily. Suppose I wanted to square the number 13. Instead of multiplying 13×13,

Numbers that add to 26		Distance from 13	Their product	Product's difference from 169
13	13	0	169	0
12	14	1	168	1
11	15	2	165	4
10	16	3	160	9
9	17	4	153	16
8	18	5	144	25

why not get an approximate answer by using two numbers that are easier to multiply but also add to 26? I chose $10 \times 16 = 160$. To get the final answer, I just added $3^2 = 9$ (since 10 and 16 are each 3 away from 13). Thus, $13^2 = 160 + 9 = 169$. Neat!

This method is diagrammed as follows:

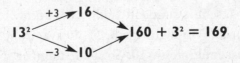

$$13^2 \quad \overset{+3}{\nearrow} 16 \quad \overset{-3}{\searrow} 10 \quad \rightarrow 160 + 3^2 = 169$$

Now let's see how this works for another square:

To square 41, subtract 1 to obtain 40 and add 1 to obtain 42. Next multiply 40 × 42. Don't panic! This is simply a 2-by-1 multiplication problem (specifically, 4 × 42) in disguise. Since 4 × 42 = 168, 40 × 42 = 1680. Almost done! All you have to add is the square of 1 (the number by which you went up and down from 41), giving you 1680 + 1 = 1681.

Can squaring a two-digit number be this easy? Yes, with this method and a little practice, it can. And it works whether you initially round down or round up. For example, let's examine 77^2, working it out both by rounding up and by rounding down:

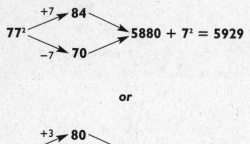

or

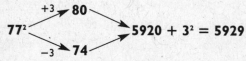

In this instance the advantage of rounding up is that you are virtually done as soon as you have completed the multiplication problem because it is simple to add 9 to a number ending in 0!

In fact, for all two-digit squares, I always round up or down to the nearest multiple of 10. So if the number to be squared ends in 6, 7, 8, or 9, round up, and if the number to be squared ends in 1, 2, 3, or 4, round down. (If the number ends in 5, you do both!) With this strategy you will add only the numbers 1, 4, 9, 16, or 25 to your first calculation.

Let's try another problem. Calculate 56^2 in your head before looking at how we did it, below:

$$56^2 \xrightarrow[-4]{+4} \begin{array}{c} 60 \\ 52 \end{array} \longrightarrow 3120 + 4^2 = 3136$$

Squaring numbers that end in 5 is even easier. Since you will always round up and down by 5, the numbers to be multiplied will both be multiples of 10. Hence, the multiplication and the addition are especially simple. We have worked out 85^2 and 35^2, below:

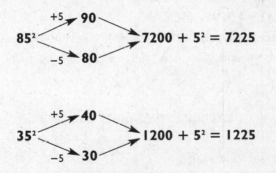

$$85^2 \xrightarrow[-5]{+5} \begin{array}{c} 90 \\ 80 \end{array} \longrightarrow 7200 + 5^2 = 7225$$

$$35^2 \xrightarrow[-5]{+5} \begin{array}{c} 40 \\ 30 \end{array} \longrightarrow 1200 + 5^2 = 1225$$

As you saw in Chapter 0, when you are squaring a number that ends in 5, rounding up and down allows you to blurt out the first part of the answer immediately and then finish it with 25. For example, if you want to compute 75^2, rounding up to 80 and down to 70 will give you "Fifty-six hundred and . . . twenty-five!"

For numbers ending in 5, you should have no trouble beating someone with a calculator, and with a little practice with the other squares, it won't be long before you can beat the calculator with any two-digit square number. Even large numbers are not to be feared. You can ask someone to give you a really big two-digit number, something in the high 90s, and it will sound as though you've chosen an impossible problem to compute. But, in fact, these are even easier because they allow you to round up to 100.

Let's say your audience gives you 96^2. Try it yourself, and then check how we did it.

Wasn't that easy? You should have rounded up by 4 to 100 and down by 4 to 92, and then multiplied 100 × 92 to get 9200. At this point you can say out loud, "Ninety-two hundred," and then finish up with "sixteen" and enjoy the applause!

EXERCISE: TWO-DIGIT SQUARES

Compute the following:

1. 14^2	2. 27^2	3. 65^2	4. 89^2	5. 98^2
6. 31^2	7. 41^2	8. 59^2	9. 26^2	10. 53^2
11. 21^2	12. 64^2	13. 42^2	14. 55^2	15. 75^2
16. 45^2	17. 84^2	18. 67^2	19. 103^2	20. 208^2

Zerah Colburn: Entertaining Calculations

One of the first lightning calculators to capitalize on his talent was Zerah Colburn (1804–1839), an American farmer's son from Vermont who learned the multiplication tables to 100 before he could even read or write. By the age of six, young Zerah's father took him on the road, where his performances generated enough capital to send him to school in Paris and London. By age eight he was internationally famous, performing lightning calculations in England, and was described in the *Annual Register* as "the most singular phenomenon in the history of the human mind that perhaps ever existed." No less than Michael Faraday and Samuel Morse admired him.

No matter where he went, Colburn met all challengers with speed and precision. He tells us in his autobiography of one set of problems he was given in New Hampshire in June 1811: "How many days and hours since the Christian Era commenced, 1811 years ago? Answered in twenty seconds: 661,015 days, 15,864,360 hours. How many seconds in eleven years? Answered in four seconds; 346,896,000." Colburn used the same techniques described in this book to compute entirely in his head problems given to him. For example, he would factor large numbers into smaller numbers and then multiply: Colburn once multiplied 21,734 × 543 by factoring 543 into 181 × 3. He then multiplied 21,734 × 181 to arrive at 3,933,854, and finally multiplied that figure by 3, for a total of 11,801,562.

As is often the case with lightning calculators, interest in Colburn's amazing skills diminished with time, and by the age of twenty he had returned to America and become a Methodist preacher. He died at a youthful thirty-five. In summarizing his skills as a lightning calculator, and the advantage such an ability affords, Colburn reflected, "True, the method . . . requires a much larger number of figures than the common Rule, but it will be remembered that pen, ink and paper cost Zerah very little when engaged in a sum."

WHY THESE TRICKS WORK

This section is presented for teachers, students, math buffs, and anyone curious as to why our methods work. Some people may find the theory as interesting as the application. Fortunately, you need not understand why our methods work in order to understand how to apply them. All magic tricks have a rational explanation behind them, and mathemagical tricks are no different. It is here that the mathemagician reveals his deepest secrets!

In this chapter on multiplication problems, the distributive law is what allows us to break down problems into their component parts. The distributive law states that for any numbers a, b, and c:

$$(b + c) \times a = (b \times a) + (c \times a)$$

That is, the outside term, a, is distributed, or separately applied, to each of the inside terms, b and c. For example, in our first mental multiplication problem of 42×7, we arrived at the answer by treating 42 as $40 + 2$, then distributing the 7 as follows:

$$42 \times 7 = (40 + 2) \times 7 = (40 \times 7) + (2 \times 7) = 280 + 14 = 294$$

You may wonder why the distributive law works in the first place. To understand it intuitively, imagine having 7 bags, each containing 42 coins, 40 of which are gold and 2 of which are silver. How many coins do you have altogether? There are two ways to arrive at the answer. In the first place, by the very definition of multiplication, there are 42×7 coins. On the other hand, there are 40×7 gold coins and 2×7 silver coins. Hence, we have $(40 \times 7) + (2 \times 7)$ coins altogether. By answering our question two ways, we have $42 \times 7 = (40 \times 7) + (2 \times 7)$. Notice that the numbers 7, 40, and 2 could be replaced by any numbers (a, b, or c) and the same logic would apply. That's why the distributive law works!

Using similar reasoning with gold, silver, and copper coins we can derive:

$$(b + c + d) \times a = (b \times a) + (c \times a) + (d \times a)$$

Hence, to do the problem 326×7, we break up 326 as 300 + 20 + 6, then distribute the 7, as follows: $326 \times 7 = (300 + 20 + 6) \times 7 = (300 \times 7) + (20 \times 7) + (6 \times 7)$, which we then add up to get our answer.

As for squaring, the following algebra justifies my method. For any numbers A and d

$$A^2 = (A + d) \times (A - d) + d^2$$

Here, A is the number being squared; d can be any number, but I choose it to be the distance from A to the nearest multiple of 10. Hence, for 77^2, I set $d = 3$ and our formula tells us that $77^2 = (77 + 3) \times (77 - 3) + 3^2 = (80 \times 74) + 9 = 5929$. The following algebraic relationship also works to explain my squaring method:

$$(z + d)^2 = z^2 + 2zd + d^2 = z(z + 2d) + d^2$$

Hence, to square 41, we set $z = 40$ and $d = 1$ to get:

$$41^2 = (40 + 1)^2 = 40 \times (40 + 2) + 1^2 = 1681$$

Similarly,

$$(z - d)^2 = z(z - 2d) + d^2$$

To find 77^2 when $z = 80$ and $d = 3$,

$$77^2 = (80 - 3)^2 = 80 \times (80 - 6) + 3^2 = 80 \times 74 + 9 = 5929$$

Chapter 3

New and Improved Products: Intermediate Multiplication

Mathemagics really gets exciting when you perform in front of an audience. I experienced my first public performance in eighth grade, at the fairly advanced age of thirteen. Many mathemagicians begin even earlier. Zerah Colburn (1804–1839), for example, reportedly could do lightning calculations before he could read or write, and he was entertaining audiences by the age of six! When I was thirteen, my algebra teacher did a problem on the board for which the answer was 108^2. Not content to stop there, I blurted out, "108 squared is simply 11,664!"

The teacher did the calculation on the board and arrived at the same answer. Looking a bit startled, she said, "Yes, that's right. How did you do it?" So I told her, "I went down 8 to 100 and up 8 to 116. I then multiplied 116 × 100, which is 11,600, and just added the square of 8, to get 11,664."

She had never seen that method before. I was thrilled. Thoughts of "Benjamin's Theorem" popped into my head. I actually believed I had discovered something new. When I finally ran across this method a few years later in a book by Martin

Gardner on recreational math, *Mathematical Carnival* (1965), it ruined my day! Still, the fact that I had discovered it for myself was very exciting to me.

You, too, can impress your friends (or teachers) with some fairly amazing mental multiplication. At the end of the last chapter you learned how to multiply a two-digit number by itself. In this chapter you will learn how to multiply two different two-digit numbers, a challenging yet more creative task. You will then try your hand—or, more accurately, your brain—at three-digit squares. You do not have to know how to do 2-by-2 multiplication problems to tackle three-digit squares, so you can learn either skill first.

2-BY-2 MULTIPLICATION PROBLEMS

When squaring two-digit numbers, the method is always the same. When multiplying two-digit numbers, however, you can use lots of different methods to arrive at the same answer. For me, this is where the fun begins.

The first method you will learn is the "addition method," which can be used to solve all 2-by-2 multiplication problems.

The Addition Method

To use the addition method to multiply any two two-digit numbers, all you need to do is perform two 2-by-1 multiplication problems and add the results together. For example:

$$
\begin{array}{r}
46 \\
\times\ \ 42\ (40 + 2) \\
\hline
\end{array}
$$

$$
\begin{array}{rr}
40 \times 46 = & 1840 \\
2 \times 46 = & +\ \ 92 \\
\hline
& 1932
\end{array}
$$

Here you break up 42 into 40 and 2, two numbers that are easy to multiply. Then you multiply 40 × 46, which is just 4 × 46 with a 0 attached, or 1840. Then you multiply 2 × 46 = 92. Finally, you add 1840 + 92 = 1932, as diagrammed above.

Here's another way to do the same problem:

$$
\begin{array}{r}
46\ (40 + 6) \\
\times\quad 42 \\
\hline
\end{array}
$$

$$
\begin{array}{rr}
40 \times 42 = & 1680 \\
6 \times 42 = & +\ 252 \\
\hline
& 1932
\end{array}
$$

The catch here is that multiplying 6 × 42 is harder to do than multiplying 2 × 46, as in the first problem. Moreover, adding 1680 + 252 is more difficult than adding 1840 + 92. So how do you decide which number to break up? I try to choose the number that will produce the easier addition problem. In most cases—but not all—you will want to break up the number with the smaller last digit because it usually produces a smaller second number for you to add.

Now try your hand at the following problems:

$$
\begin{array}{rr}
& 48 \\
\times & 73\ (70 + 3) \\
\hline
\end{array}
$$

$$
\begin{array}{rr}
70 \times 48 = & 3360 \\
3 \times 48 = & +\ 144 \\
\hline
& 3504
\end{array}
$$

$$
\begin{array}{rr}
& 81\ (80 + 1) \\
\times & 59 \\
\hline
\end{array}
$$

$$
\begin{array}{rr}
80 \times 59 = & 4720 \\
1 \times 59 = & +\ 59 \\
\hline
& 4779
\end{array}
$$

The last problem illustrates why numbers that end in 1 are especially attractive to break up. If both numbers end in the same digit, you should break up the larger number as illustrated below:

$$
\begin{array}{r}
84\ (80 + 4) \\
\times\ \ 34 \\
\hline
80 \times 34 = \quad 2720 \\
4 \times 34 = + \ 136 \\
\hline
2856
\end{array}
$$

If one number is much larger than the other, it often pays to break up the larger number, even if it has a larger last digit. You will see what I mean when you try the following problem two different ways:

$$
\begin{array}{r}
74\ (70 + 4) \\
\times\ 13 \\
\hline
70 \times 13 = \quad 910 \\
4 \times 13 = + \ 52 \\
\hline
962
\end{array}
\qquad\qquad
\begin{array}{r}
74 \\
\times\ 13\ (10 + 3) \\
\hline
10 \times 74 = \quad 740 \\
3 \times 74 = + \ 222 \\
\hline
962
\end{array}
$$

Did you find the first method to be faster than the second? I did. Here's another exception to the rule of breaking up the number with the smaller last digit. When you multiply a number in the fifties by an even number, you'll want to break up the number in the fifties:

$$
\begin{array}{r}
84 \\
\times\ 59\ (50 + 9) \\
\hline
50 \times 84 = \quad 4200 \\
9 \times 84 = + \ 756 \\
\hline
4956
\end{array}
$$

The last digit of the number 84 is smaller than the last digit of 59, but if you break up 59, your product will be a multiple of 100, just as 4200 is, in the example above. This makes the subsequent addition problem much easier.

Now try an easy problem of a different sort:

$$
\begin{array}{r}
42 \\
\times\ 11\ (10 + 1) \\
\hline
10 \times 42 = \quad 420 \\
1 \times 42 = + \ 42 \\
\hline
462
\end{array}
$$

Though the calculation above is pretty simple, there is an even *easier* and *faster* way to multiply any two-digit number by 11. This is mathemagics at its best: you won't believe your eyes when you see it (unless you remember it from Chapter 0)!

Here's how it works. Suppose you have a two-digit number whose digits add up to 9 or less. To multiply this number by 11, merely add the two digits together and insert the total between the original two digits. For example, to do 42 × 11, first do 4 + 2 = 6. If you place the 6 between the 4 and the 2, you get 462, the answer to the problem!

$$
\begin{array}{ccc}
42 & 4\_\_2 & = 462 \\
\times\ 11 & 6 &
\end{array}
$$

Try 54 × 11 by this method.

$$
\begin{array}{ccc}
54 & 5\_\_4 & = 594 \\
\times\ 11 & 9 &
\end{array}
$$

What could be simpler? All you had to do was place the 9 between the 5 and the 4 to give you the final answer of 594.

You may wonder what happens when the two numbers add up to a number larger than 9. In such cases, increase the tens digit by 1, then insert the last digit of the sum between the two numbers, as before. For example, when multiplying 76 × 11, 7 + 6 = 13, you increase the 7 in 76 to 8, and then insert the 3 between the 8 and the 6, giving you 836, the final answer. See the diagrammed answer below:

$$76 \qquad 7\ \underline{\ 6} \qquad = 836$$
$$\underline{\times 11} \qquad 1\ 3$$

Your turn. Try 68 × 11.

$$68 \qquad 6\ \underline{\ 8} \qquad = 748$$
$$\underline{\times 11} \qquad 1\ 4$$

Once you get the hang of this trick, you will never multiply by 11 any other way. Try a few more problems and then check your answers at the back of the book.

EXERCISE: MULTIPLYING BY 11

1. $$35$$
   $$\underline{\times 11}$$

2. $$48$$
   $$\underline{\times 11}$$

3. $$94$$
   $$\underline{\times 11}$$

Returning to the addition method, the next problem is a real challenge the first time you try it. Try solving 89 × 72 in your head, looking back at the problem if necessary. If you have to start over a couple of times, that's okay.

$$
\begin{array}{r}
89 \\
\times \;\;\; 72 \;(70+2) \\
\end{array}
$$

$$
\begin{array}{rr}
70 \times 89 = & 6230 \\
2 \times 89 = & +\;178 \\
\hline
& 6408 \\
\end{array}
$$

If you got the right answer the first or second time, pat yourself on the back. The 2-by-2 multiplication problems really do not get any tougher than this. If you did not get the answer right away, don't worry. In the next two sections, I'll teach you some much easier strategies for dealing with problems like this. But before you read on, practice the addition method on the following multiplication problems.

EXERCISE: 2-BY-2 ADDITION-METHOD
MULTIPLICATION PROBLEMS

1. $\begin{array}{r} 31 \\ \times 41 \end{array}$	2. $\begin{array}{r} 27 \\ \times 18 \end{array}$	3. $\begin{array}{r} 59 \\ \times 26 \end{array}$	4. $\begin{array}{r} 53 \\ \times 58 \end{array}$
5. $\begin{array}{r} 77 \\ \times 43 \end{array}$	6. $\begin{array}{r} 23 \\ \times 84 \end{array}$	7. $\begin{array}{r} 62 \\ \times 94 \end{array}$	8. $\begin{array}{r} 88 \\ \times 76 \end{array}$
9. $\begin{array}{r} 92 \\ \times 35 \end{array}$	10. $\begin{array}{r} 34 \\ \times 11 \end{array}$	11. $\begin{array}{r} 85 \\ \times 11 \end{array}$	

The Subtraction Method

The subtraction method really comes in handy when one of the numbers you want to multiply ends in 8 or 9. The following problem illustrates what I mean:

$$
\begin{array}{r}
59\,(60-1) \\
\times\quad 17 \\
\hline
\end{array}
$$

$$
\begin{array}{rr}
60 \times 17 = & 1020 \\
-1 \times 17 = & -\quad 17 \\
\hline
& 1003
\end{array}
$$

Although most people find addition easier than subtraction, it is usually easier to subtract a small number than to add a big number. (If we had done this problem by the addition method, we would have added $850 + 153 = 1003$.)

Now let's do the challenging problem from the end of the last section:

$$
\begin{array}{r}
89\,(90-1) \\
\times\quad 72 \\
\hline
\end{array}
$$

$$
\begin{array}{rr}
90 \times 72 = & 6480 \\
-1 \times 72 = & -\quad 72 \\
\hline
& 6408
\end{array}
$$

Wasn't that a whole lot easier? Now, here's a problem where one number ends in 8:

$$
\begin{array}{r}
88\,(90-2) \\
\times\quad 23 \\
\hline
\end{array}
$$

$$
\begin{array}{rr}
90 \times 23 = & 2070 \\
-2 \times 23 = & -\quad 46 \\
\hline
& 2024
\end{array}
$$

In this case you should treat 88 as $90 - 2$, then multiply $90 \times 23 = 2070$. But you multiplied by too much. How much? By 2×23, or 46 too much. So subtract 46 from 2070 to arrive at 2024, the final answer.

I want to emphasize here that it is important to work out these problems in your head and not simply look to see how we did it in the diagram. Go through them and say the steps to yourself or even out loud to reinforce your thoughts.

Not only do I use the subtraction method with numbers that end in 8 or 9, but also for numbers in the high 90s because 100 is such a convenient number to multiply. For example, if someone asked me to multiply 96 × 73, I would immediately round up 96 to 100:

$$
\begin{array}{r}
96\,(100 - 4) \\
\times\ \ 73 \\
\end{array}
$$

$$
\begin{aligned}
100 \times 73 &= 7300 \\
-4 \times 73 &= -\ 292 \\
\hline
&\ \ 7008
\end{aligned}
$$

When the subtraction component of a multiplication problem requires you to borrow a number, using complements (as we learned in Chapter 1) can help you arrive at the answer more quickly. You'll see what I mean as you work your way through the problems below. For example, subtract 340 − 78. We know the answer will be in the 200s. The difference between 40 and 78 is 38. Now take the complement of 38 to get 62. And that's the answer, 262!

$$
\begin{array}{r}
340 \\
-\ \ 78 \\
\hline
262
\end{array}
\qquad
\begin{aligned}
78 - 40 &= 38 \\
\text{Complement of } 38 &= 62
\end{aligned}
$$

Now let's try another problem:

$$88\ (90 - 2)$$
$$\times\ \ 76$$

$$90 \times 76 =\ \ \ 6840$$
$$-2 \times 76 = -\ \ 152$$

There are two ways to perform the subtraction component of this problem. The "long" way subtracts 200 and adds back 48:

$$6840 - 152\ \ \ =\ \ \ 6640 + 48\ \ \ =\ \ \ 6688$$
$$\text{(first subtract 200)}\ \ \ \ \ \ \ \ \ \text{(then add 48)}$$

The *short* way is to realize that the answer will be 66 hundred and *something*. To determine *something,* we subtract $52 - 40 = 12$ and then find the complement of 12, which is 88. Hence the answer is 6688.

Try this one.

$$67$$
$$\times\ \ \ 59\ (60 - 1)$$

$$60 \times 67 =\ \ \ 4020$$
$$-1 \times 67 = -\ \ \ 67$$
$$3953$$

Again, you can see that the answer will be 3900 and something. Because $67 - 20 = 47$, the complement 53 means the answer is 3953.

As you may have realized, you can use this method with any subtraction problem that requires you to borrow a number, not just those that are part of a multiplication problem. All of this is further proof, if you need it, that complements are a very powerful tool in mathemagics. Master this technique and, pretty soon, people will be complimenting you!

EXERCISE: 2-BY-2 SUBTRACTION-METHOD
MULTIPLICATION PROBLEMS

1.	29	2.	98	3.	47	4.	68
	× 45		× 43		× 59		× 38

5.	96	6.	79	7.	37	8.	87
	× 29		× 54		× 19		× 22

9.	85	10.	57	11.	88
	× 38		× 39		× 49

The Factoring Method

The factoring method is my favorite method of multiplying two-digit numbers since it involves no addition or subtraction at all. You use it when one of the numbers in a two-digit multiplication problem can be factored into one-digit numbers.

To factor a number means to break it down into one-digit numbers that, when multiplied together, give the original number. For example, the number 24 can be factored into 8 × 3 or 6 × 4. (It can also be factored into 12 × 2, but we prefer to use only single-digit factors.)

Here are some other examples of factored numbers:

$$42 = 7 \times 6$$
$$63 = 9 \times 7$$
$$84 = 7 \times 6 \times 2 \text{ or } 7 \times 4 \times 3$$

To see how factoring makes multiplication easier, consider the following problem:

$$46$$
$$\underline{\times\ 42} = 7 \times 6$$

Previously we solved this problem by multiplying 46 × 40 and 46 × 2 and adding the products together. To use the factoring method, treat 42 as 7 × 6 and begin by multiplying 46 × 7, which is 322. Then multiply 322 × 6 for the final answer of 1932. You already know how to do 2-by-1 and 3-by-1 multiplication problems, so this should not be too hard:

$$46 \times 42 = 46 \times (7 \times 6) = (46 \times 7) \times 6 = 322 \times 6 = 1932$$

Of course, this problem could also have been solved by reversing the factors of 42:

$$46 \times 42 = 46 \times (6 \times 7) = (46 \times 6) \times 7 = 276 \times 7 = 1932$$

In this case, it is easier to multiply 322 × 6 than it is to multiply 276 × 7. In most cases, I like to use the larger factor in solving the initial 2-by-1 problem and to reserve the smaller factor for the 3-by-1 component of the problem.

Factoring results in a 2-by-2 multiplication problem being simplified to an easier 3-by-1 (or sometimes 2-by-1) multiplication problem. The advantage of the factoring method in mental calculation is you do not have to hold much in memory. Let's look at another example, 75 × 63:

$$75 \times 63 = 75 \times (9 \times 7) = (75 \times 9) \times 7 = 675 \times 7 = 4725$$

As before, you simplify this 2-by-2 problem by factoring 63 into 9 × 7 and then multiplying 75 by these factors. (By the way, the reason we can shift parentheses in the second step is the *associative law* of multiplication.)

$$63 \times 75 = 63 \times (5 \times 5 \times 3) = (63 \times 5) \times 5 \times 3$$
$$= 315 \times 5 \times 3 = 1575 \times 3 = 4725$$

Try the following problem for practice:

$$57 \times 24 = 57 \times 8 \times 3 = 456 \times 3 = 1368$$

You could have factored 24 as 6 × 4 for another easy computation:

$$57 \times 24 = 57 \times 6 \times 4 = 342 \times 4 = 1368$$

Compare this approach with the addition method:

	57	*or*
	× 24 (20 + 4)	
20 × 57 =	1140	
4 × 57 =	+ 228	
	1368	

	57 (50 + 7)
	× 24
50 × 24 =	1200
7 × 24 =	+ 168
	1368

With the addition method, you have to perform two 2-by-1 problems and then add. With the factoring method, you have just two multiplication problems: a 2-by-1 and a 3-by-1, and then you are done. The factoring method is usually easier on your memory.

Remember that challenging multiplication problem earlier in this chapter? Here it is again:

$$89 \times 72$$

We tackled that problem easily enough with the subtraction method, but factoring works even faster:

$$89 \times 72 = 89 \times 9 \times 8 = 801 \times 8 = 6408$$

The problem is especially easy because of the 0 in the middle of 801. Our next example illustrates that it sometimes pays to factor the numbers in an order that exploits this situation. Let's look at two ways of computing 67×42:

$$67 \times 42 = 67 \times 7 \times 6 = 469 \times 6 = 2814$$
$$67 \times 42 = 67 \times 6 \times 7 = 402 \times 7 = 2814$$

Ordinarily you should factor 42 into 7×6, as in the first example, following the rule of using the larger factor first. But the problem is easier to solve if you factor 42 into 6×7 because it creates a number with a 0 in the center, which is easier to multiply. I call such numbers *friendly products*.

Look for the friendly product in the problem done two ways below:

$$43 \times 56 = 43 \times 8 \times 7 = 344 \times 7 = 2408$$
$$43 \times 56 = 43 \times 7 \times 8 = 301 \times 8 = 2408$$

Did you think the second way was easier?

When using the factoring method, it pays to find friendly

products whenever you can. The following list should help. I don't expect you to memorize it so much as to familiarize yourself with it. With practice you will be able to nose out friendly products more often, and the list will become more meaningful.

Numbers with Friendly Products

12: $12 \times 9 = 108$

13: $13 \times 8 = 104$

15: $15 \times 7 = 105$

17: $17 \times 6 = 102$

18: $18 \times 6 = 108$

21: $21 \times 5 = 105$

23: $23 \times 9 = 207$

25: $25 \times 4 = 100, 25 \times 8 = 200$

26: $26 \times 4 = 104, 26 \times 8 = 208$

27: $27 \times 4 = 108$

29: $29 \times 7 = 203$

34: $34 \times 3 = 102, 34 \times 6 = 204, 34 \times 9 = 306$

35: $35 \times 3 = 105$

36: $36 \times 3 = 108$

38: $38 \times 8 = 304$

41: $41 \times 5 = 205$

43: $43 \times 7 = 301$

44: $44 \times 7 = 308$

45: $45 \times 9 = 405$

51: $51 \times 2 = 102, 51 \times 4 = 204, 51 \times 6 = 306, 51 \times 8 = 408$

52: $52 \times 2 = 104, 52 \times 4 = 208$

53: $53 \times 2 = 106$

54: $54 \times 2 = 108$

56: $56 \times 9 = 504$

61: $61 \times 5 = 305$

63: $63 \times 8 = 504$

67: $67 \times 3 = 201$, $67 \times 6 = 402$, $67 \times 9 = 603$
68: $68 \times 3 = 204$, $68 \times 6 = 408$
69: $69 \times 3 = 207$
72: $72 \times 7 = 504$
76: $76 \times 4 = 304$, $76 \times 8 = 608$
77: $77 \times 4 = 308$
78: $78 \times 9 = 702$
81: $81 \times 5 = 405$
84: $84 \times 6 = 504$
86: $86 \times 7 = 802$
88: $88 \times 8 = 704$
89: $89 \times 9 = 801$

Previously in this chapter you learned how easy it is to multiply numbers by 11. It usually pays to use the factoring method when one of the numbers is a multiple of 11, as in the examples below:

$$52 \times 33 = 52 \times 11 \times 3 = 572 \times 3 = 1716$$
$$83 \times 66 = 83 \times 11 \times 6 = 913 \times 6 = 5478$$

EXERCISE: 2-BY-2 FACTORING-METHOD

MULTIPLICATION PROBLEMS

---

1. 27	2. 86	3. 57	4. 81
× 14	× 28	× 14	× 48

5. 56	6. 83	7. 72	8. 85
× 29	× 18	× 17	× 42

9. 33	10. 62	11. 45	12. 48
× 16	× 77	× 36	× 37

APPROACHING MULTIPLICATION CREATIVELY

I mentioned at the beginning of the chapter that multiplication problems are fun because they can be solved any number of ways. Now that you know what I mean, let's apply all three methods explained in this chapter to a single problem, 73 × 49. We'll begin by using the addition method:

$$
\begin{array}{r}
73\,(70 + 3) \\
\times\quad 49 \\
\hline
70 \times 49 = \quad 3430 \\
3 \times 49 = +\;\; 147 \\
\hline
3577
\end{array}
$$

Now try the subtraction method:

$$
\begin{array}{r}
73 \\
\times\quad 49\,(50 - 1) \\
\hline
50 \times 73 = \quad 3650 \\
-1 \times 73 = -\;\; 73 \\
\hline
3577
\end{array}
$$

Note that the last two digits of the subtraction could be obtained by adding 50 + (complement of 73) = 50 + 27 = 77 or by simply taking the complement of (73 − 50) = complement of 23 = 77.

Finally, try the factoring method:

$$73 \times 49 = 73 \times 7 \times 7 = 511 \times 7 = 3577$$

Congratulations! You have mastered 2-by-2 multiplication and now have all the basic skills you need to be a fast mental calculator. All you need to become a lightning calculator is more practice!

EXERCISE: 2-BY-2 GENERAL MULTIPLICATION—ANYTHING GOES!

Many of the following exercises can be solved by more than one method. Try computing them in as many ways as you can think of, then check your answers and computations at the back of the book. Our answers suggest various ways the problem can be mathemagically solved, starting with what I think is the easiest method.

1.	2.	3.	4.	5.
53	81	73	89	77
× 39	× 57	× 18	× 55	× 36

6.	7.	8.	9.	10.
92	87	67	56	59
× 53	× 87	× 58	× 37	× 21

The following 2-by-2s occur as subproblems to problems appearing later when we do 3-by-2s, 3-by-3s, and 5-by-5s. You can do these problems now for practice, and refer back to them when they are used in the larger problems.

11.	12.	13.	14.	15.
37	57	38	43	43
× 72	× 73	× 63	× 76	× 75

16.	17.	18.	19.	20.
74	61	36	54	53
× 62	× 37	× 41	× 53	× 53

21.	22.	23.	24.	25.
83	91	52	29	41
× 58	× 46	× 47	× 26	× 15

26.	65	27.	34	28.	69	29.	95	30.	65
	× 19		× 27		× 78		× 81		× 47

31.	65	32.	95	33.	41
	× 69		× 26		× 93

THREE-DIGIT SQUARES

Squaring three-digit numbers is an impressive feat of mental prestidigitation. Just as you square two-digit numbers by rounding up or down to the nearest multiple of 10, to square three-digit numbers, you round up or down to the nearest multiple of 100. Take 193:

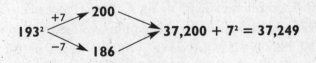

By rounding up to 200 and down to 186, you've transformed a 3-by-3 multiplication problem into a far simpler 3-by-1 problem. After all, 200×186 is just $2 \times 186 = 372$ with two zeros attached. Almost done! Now all you have to add is $7^2 = 49$ to arrive at 37,249.

Now try squaring 706:

Rounding *down* by 6 to 700 requires you to round up by 6 to 712. Since 712 × 7 = 4984 (a simple 3-by-1 problem), 712 × 700 = 498,400. After adding 6^2 = 36, you arrive at 498,436.

These last problems are not terribly hard because there is no real addition involved. Moreover, you know the answers to 6^2 and 7^2 by heart. Squaring a number that's farther away from a multiple of 100 is a tougher proposition. Try your hand at 314^2:

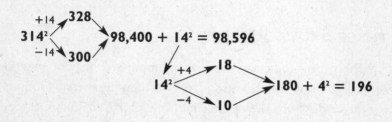

For this three-digit square, go down 14 to 300 and up 14 to 328, then multiply 328 × 3 = 984. Tack on two 0s to arrive at 98,400. Then add the square of 14. If 14^2 = 196 comes to you in a flash (through memory or calculation), you're in good shape. Just add 98,400 + 196 to arrive at 98,596. If you need time to compute 14^2, repeat the number 98,400 to yourself a few times before you go on. (Otherwise you might compute 14^2 = 196 and forget what number to add it to.)

The farther away you get from a multiple of 100, the more difficult squaring a three-digit number becomes. Try 529^2:

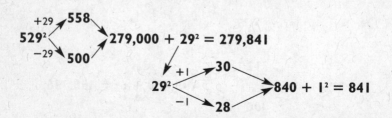

If you have an audience you want to impress, you can say 279,000 out loud before you compute 29^2. But this will not work for every problem. For instance, try squaring 636:

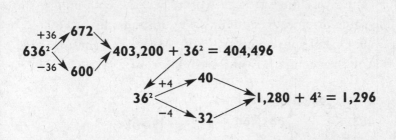

Now your brain is really working, right? The key here is to repeat 403,200 to yourself several times. Then square 36 to get 1,296 in the usual way. The hard part comes in adding 1,296 to 403,200. Do it one digit at a time, left to right, to arrive at your answer of 404,496. Take my word that as you become more familiar with two-digit squares, these three-digit problems get easier.

Here's an even tougher problem, 863^2:

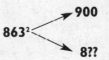

The first problem is deciding what numbers to multiply together. Clearly one of the numbers will be 900, and the other number will be in the 800s. But what number? You can compute it two ways:

1. The hard way: the difference between 863 and 900 is 37 (the complement of 63). Subtract 37 from 863 to arrive at 826.

2. The easy way: double the number 63 to get 126, and take the last two digits to give you 826.

Here's why the easy way works. Because both numbers are the same distance from 863, their sum must be twice 863, or 1726. One of your numbers is 900, so the other must be 826. You then compute the problem like this:

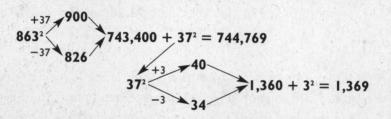

If you find it impossible to remember 743,400 after squaring 37, fear not. In a later chapter, you will learn a memory system that will make remembering such numbers much easier.

Try your hand at squaring 359, the hardest problem yet:

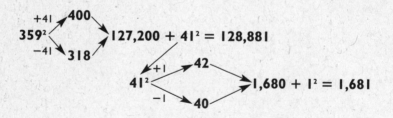

To obtain 318, either subtract 41 (the complement of 59) from 359, or multiply $2 \times 59 = 118$ and use the last two digits. Next multiply $400 \times 318 = 127,200$. Adding 41^2, or 1,681, gives you 128,881. Whew! They don't get much harder than that! If you got it right the first time, take a bow!

Let's finish this section with a big problem that is easy to do, 987^2:

What's Behind Door Number 1?

The mathematical chestnut of 1991 that got everyone hopping mad was an article in *Parade* magazine by Marilyn vos Savant, the woman listed by the *Guinness Book of World Records* as having the world's highest IQ. The paradox has come to be known as the Monty Hall problem, and it goes like this.

You are a contestant on *Let's Make a Deal*. Monty Hall allows you to pick one of three doors; behind one of these doors is the big prize, behind the other two are goats. You pick Door Number 2. But before Monty reveals the prize of your choice, he shows you what you didn't pick behind Door Number 3. It's a goat. Now, in his tantalizing way, Monty gives you another choice. Do you want to stick with Door Number 2, or do you want to risk a chance to see what's behind Door Number 1? What should you do? Assuming that Monty is only going to reveal where the big prize is not, he will always open one of the consolation doors. This leaves two doors, one with the big prize and the other with another consolation. The odds are now 50-50 for your choice, right?

Wrong! The odds that you chose correctly the first time remain 1 in 3. The probability that the big prize is behind the other door increases to 2 in 3 because the probability must add to 1.

Thus, by switching doors, you double the odds of winning! (The problem assumes that Monty will always give a player the option to switch, that he will always reveal a nonwinning door, and that when your first pick is correct he will choose a nonwinning door at random.) Think of playing the game with ten doors and after your pick he reveals eight other nonwinning doors. Here, your instincts would probably tell you to switch. People confuse this problem for a variant: if Monty Hall does not know where the grand prize is, and reveals Door Number 3, which happens to contain a goat (though it might have contained the prize), then Door Number 1 has a 50 percent chance of being correct. This result is so counterintuitive that Marilyn vos Savant received piles of letters, many from scientists and even mathematicians, telling her she shouldn't write about math. They were all wrong.

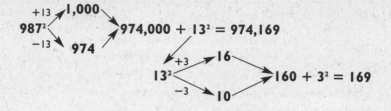

EXERCISE: THREE-DIGIT SQUARES

1. 409^2 2. 805^2 3. 217^2 4. 896^2

5. 345^2 6. 346^2 7. 276^2 8. 682^2

9. 431^2 10. 781^2 11. 975^2

CUBING

We end this chapter with a new method for cubing two-digit numbers. (Recall that the cube of a number is that number multiplied by itself twice. For example, 5 cubed—denoted 5^3—is equal to $5 \times 5 \times 5 = 125$.) As you will see, this is not much harder than multiplying two-digit numbers. The method is based on the algebraic observation that

$$A^3 = (A - d)A(A + d) + d^2A$$

where d is any number. Just like with squaring two-digit numbers, I choose d to be the distance to the nearest multiple of ten. For example, when squaring 13, we let $d = 3$, resulting in:

$$13^3 = (10 \times 13 \times 16) + (3^2 \times 13)$$

Since $13 \times 16 = 13 \times 4 \times 4 = 52 \times 4 = 208$, and $9 \times 13 = 117$, we have

$$13^3 = 2080 + 117 = 2197$$

How about the cube of 35? Letting $d = 5$, we get

$$35^3 = (30 \times 35 \times 40) + (5^2 \times 35)$$

Since $30 \times 35 \times 40 = 30 \times 1,400 = 42,000$ and $35 \times 5 \times 5 = 175 \times 5 = 875$, we get

$$35^3 = 42,000 + 875 = 42,875$$

When cubing 49, we let $d = 1$ in order to round up to 50. Here

$$49^3 = (48 \times 49 \times 50) + (1^2 \times 49)$$

We can solve 48×49 by the factoring method, but for this sort of problem, I prefer to use the close-together method, which will be described in Chapter 8. (Go ahead and look at it now, if you'd like!) Using that method, we get $48 \times 49 = (50 \times 47) + (1 \times 2) = 2352$. Multiplying this by 50, we get 117,600 and therefore

$$49^3 = 117,600 + 49 = 117,649$$

Here's a larger one. Try the cube of 92.

$$92^3 = (90 \times 92 \times 94) + (2^2 \times 92)$$

If you are fast at squaring two-digit numbers, then you could solve $92 \times 94 = 93^2 - 1 = 8648$, or you could use the close-together method, resulting in $92 \times 94 = (90 \times 96) + (2 \times 4) = 8648$. Multiplying this by 9 (as described in the beginning of

Chapter 8), we obtain $9 \times (8,600 + 48) = 77,400 + 432 = 77,832$, and therefore $90 \times 92 \times 94 = 778,320$. Since $4 \times 92 = 368$, we get

$$92^3 = 778,320 + 368 = 778,688$$

We note that when the close-together method is used for the multiplication problems that arise when cubing two-digit numbers, the small product being added will either be (depending on whether d = 1, 2, 3, 4, or 5) $1 \times 2 = 2$, $2 \times 4 = 8$, $3 \times 6 = 18$, $4 \times 8 = 32$, or $5 \times 10 = 50$. Let's finish with the cube of 96.

$$96^3 = (92 \times 96 \times 100) + (4^2 \times 96)$$

The product $92 \times 96 = 8,832$ can be done many different ways. To celebrate the end of this chapter, let's do some of them. I'll start with what I consider to be the hardest way, and end with what I consider the easiest way. By the addition method, $(90 + 2) \times 96 = 8,640 + 192 = 8,832$; by the subtraction method, $92 \times (100 - 4) = 9,200 - 368 = 8,832$; by the factoring method, $92 \times 6 \times 4 \times 4 = 552 \times 4 \times 4 = 2,208 \times 4 = 8,832$; by squaring, $94^2 - 2^2 = 8,836 - 4 = 8,832$; by the close-together method with a base of 90, $(90 \times 98) + (2 \times 6) = 8,820 + 12 = 8,832$; and by the close-together method with a base of 100, $(100 \times 88) + (-8 \times -4) = 8,800 + 32 = 8,832$.

The product $4^2 \times 96 = 1,536$ can also be done a few different ways, such as $96 \times 4 \times 4 = 384 \times 4 = 1,536$, or as $16 \times (100 - 4) = 1,600 - 64 = 1,536$. Finally, since $8,832 \times 100 = 883,200$, we have

$$96^3 = 883,200 + 1,536 = 884,736$$

EXERCISE: TWO-DIGIT CUBES

1. 12^3 2. 17^3 3. 21^3 4. 28^3 5. 33^3

6. 39^3 7. 40^3 8. 44^3 9. 52^3 10. 56^3

11. 65^3 12. 71^3 13. 78^3 14. 85^3 15. 87^3

16. 99^3

Chapter 4

Divide and Conquer: Mental Division

Mental division is a particularly handy skill to have, both in business and in daily life. How many times a week are you confronted with situations that call on you to evenly divide things, such as a check at a restaurant? This same skill comes in handy when you want to figure out the cost per unit of a case of dog food on sale, or to split the pot in poker, or to figure out how many gallons of gas you can buy with a $20 bill. The ability to divide in your head can save you the inconvenience of having to pull out a calculator every time you need to compute something.

With mental division, the left-to-right method of calculation comes into its own. This is the same method we all learned in school, so you will be doing what comes naturally. I remember as a kid thinking that this left-to-right method of division is the way all arithmetic should be done. I have often speculated that if the schools could have figured out a way to teach division right-to-left, they probably would have done so!

ONE-DIGIT DIVISION

The first step when dividing mentally is to figure out how many digits will be in your answer. To see what I mean, try on the following problem for size:

$$179 \div 7$$

To solve $179 \div 7$, we're looking for a number, Q, such that 7 times Q is 179. Now, since 179 lies between $7 \times 10 = 70$ and $7 \times 100 = 700$, Q must lie between 10 and 100, which means our answer must be a two-digit number. Knowing that, we first determine the largest multiple of 10 that can be multiplied by 7 whose answer is below 179. We know that $7 \times 20 = 140$ and $7 \times 30 = 210$, so our answer must be in the twenties. At this point we can actually *say* the number 20 since that part of our answer will certainly not change. Next we subtract $179 - 140 = 39$. Our problem has now been reduced to the division problem $39 \div 7$. Since $7 \times 5 = 35$, which is 4 away from 39, we have the rest of our answer, namely 5 with a remainder of 4, or 5 and $\frac{4}{7}$. Altogether, we have our answer, 25 with a remainder of 4, or if you prefer, $25\frac{4}{7}$. Here's what the process looks like:

$$
\begin{array}{r}
25 \\
7{\overline{)179}} \\
-140 \\
\hline
39 \\
-\ 35 \\
\hline
4 \ \leftarrow \text{remainder}
\end{array}
$$

Answer: 25 with a remainder of 4, or $25\dfrac{4}{7}$

Let's try a similar division problem using the same method of mental computation:

$$675 \div 8$$

As before, since 675 falls between $8 \times 10 = 80$ and $8 \times 100 = 800$, your answer must be below 100 and therefore is a two-digit number. To divide 8 into 675, notice that $8 \times 80 = 640$ and $8 \times 90 = 720$. Therefore, your answer is 80 something. But what is that "something"? To find out, subtract 640 from 675 for a remainder of 35. After saying the 80, our problem has been reduced to $35 \div 8$. Since $8 \times 4 = 32$, the final answer is 84 with a remainder of 3, or $84\frac{3}{8}$. We illustrate this problem as follows:

$$
\begin{array}{r}
84 \\
8\overline{)675} \\
-640 \\
\hline
35 \\
-\ 32 \\
\hline
3 \ \longleftarrow \text{remainder}
\end{array}
$$

Answer: 84 with a remainder of 3, or $84\dfrac{3}{8}$

Like most mental calculations, division can be thought of as a process of simplification. The more you calculate, the simpler the problem becomes. What began as $675 \div 8$ was simplified to a smaller problem, $35 \div 8$.

Now let's try a division problem that results in a three-digit answer:

$$947 \div 4$$

This time, your answer will have three digits because 947 falls between $4 \times 100 = 400$ and $4 \times 1000 = 4000$. Thus we must first find the largest multiple of 100 that can be *squeezed* into 947. Since $4 \times 200 = 800$, our answer is definitely in the 200s, so go ahead and say it! Subtracting 800 from 947 gives us our new division problem, $147 \div 4$. Since $4 \times 30 = 120$, we can now say the 30. After subtracting 120 from 147, we compute $27 \div 4$ to obtain the rest of the answer: 6 with a remainder of 3. Altogether, we have 236 with a remainder of 3, or $236\frac{3}{4}$.

$$
\begin{array}{r}
236 \\
4\overline{)947} \\
-800 \\
\hline
147 \\
-120 \\
\hline
27 \\
-\ 24 \\
\hline
3 \ \longleftarrow \text{remainder}
\end{array}
$$

Answer: $236\dfrac{3}{4}$

The process is just as easy when dividing a one-digit number into a four-digit number, as in our next example.

$2196 \div 5$

Here the answer will be in the hundreds because 2196 is between $5 \times 100 = 500$ and $5 \times 1000 = 5000$. After subtracting $5 \times 400 = 2000$, we can say the 400, and our problem has reduced to $196 \div 5$, which can be solved as in the previous examples.

$$439$$
$$5\overline{)2196}$$
$$-2000$$
$$196$$
$$-\ 150$$
$$46$$
$$-\ \ 45$$
$$1$$

Answer: $439\frac{1}{5}$

Actually, there is a much easier way to solve this last problem. We can simplify our problem by doubling both numbers. Since $2196 \times 2 = 4392$, we have $2196 \div 5 = 4392 \div 10 = 439.2$ or $439\frac{2}{10}$. We'll see more division shortcuts in the next section.

EXERCISE: ONE-DIGIT DIVISION

1. $318 \div 9$ 2. $726 \div 5$ 3. $428 \div 7$
4. $289 \div 8$ 5. $1328 \div 3$ 6. $2782 \div 4$

THE RULE OF "THUMB"

When dividing in your head instead of on paper, you may find it difficult to remember parts of the answer as you continue to calculate. One option, as you've seen, is to say the answer out loud as you go. But for greater dramatic effect, you may prefer, as I do, to hold the answer on your fingers and say it all together at the end. In that case, you may run into problems remembering digits greater than five if, like most of us, you have only five fingers on each hand. The solution is to use a special technique, based on sign language, which I call the Rule of "Thumb." It is most effective for remembering three-digit and greater numbers.

This technique not only is useful in this chapter but also will come in handy (pardon the pun) in subsequent chapters dealing with larger problems and longer numbers to remember.

You already know that to represent numbers 0 through 5, all you have to do is raise the equivalent number of fingers on your hand. When you get your thumb involved, it's just as easy to represent numbers 6 through 9. Here are the Rules of "Thumb":

- To hold on to 6, place your thumb on top of your pinky.
- To hold on to 7, place your thumb on top of your ring finger.
- To hold on to 8, place your thumb on top of your middle finger.
- To hold on to 9, place your thumb on top of your index finger.

With three-digit answers, hold the hundreds digit on your left hand and the tens digit on your right. When you get to the ones digit, you've reached the end of the problem (except for a possible remainder). Now say the number on your left hand, the number on your right hand, the one-digit you've just computed, and the remainder (in your head). Presto—you've said your answer!

For practice, try to compute the following four-digit division problem:

$$4579 \div 6$$

$$
\begin{array}{r}
763 \\
6\overline{)4579} \\
-4200 \\
\hline
379 \\
-\ 360 \\
\hline
19 \\
-\ 18 \\
\hline
1
\end{array}
$$

Answer: 763 $\frac{1}{6}$

In using the Rule of "Thumb" to remember the answer, you'll hold the 7 on your left hand by placing your thumb and ring finger together and the 6 on your right hand by placing your thumb and little finger together. Once you've calculated the ones digit (which is 3) and the remainder (which is 1), you can "read" the final answer off your hands from left to right: "seven . . . six . . . three with a remainder of one, or one-sixth."

Some four-digit division problems yield four-digit answers. In that case, since you have only two hands, you will have to say the thousands digit of the answer out loud and use the rule of thumb to remember the rest of the answer. For example:

$$8352 \div 3$$

```
        2784
    3)8352
     -6000
      2352
     -2100
       252
     - 240
        12
     -  12
         0
```

Answer: 2784

For this problem, you divide 3 into 8 to get your thousands digit of 2, say "two thousand" out loud, then divide 3 into 2352 in the usual way.

TWO-DIGIT DIVISION

This section assumes you already have mastered the art of dividing by a one-digit number. Naturally, division problems become harder as the number you divide by gets larger. Fortunately, I have some magic up my sleeve to make your life easier.

Let's start with a relatively easy problem first:

$$597 \div 14$$

Since 597 lies between 14×10 and 14×100, the answer (also called the quotient) lies between 10 and 100. To determine the answer, your first step is to ask how many times 14 goes into 590. Because $14 \times 40 = 560$, you know that the answer is 40 something, and so you can say "forty" out loud.

Next, subtract 560 from 597, which is 37 and reduces your problem to dividing 14 into 37. Since $14 \times 2 = 28$, your answer is 42. Subtracting 28 from 37 leaves you a remainder of 9. The process of deriving the solution to this problem may be illustrated as follows:

$$
\begin{array}{r}
42 \\
14\overline{)597} \\
-560 \\
\hline
37 \\
-28 \\
\hline
9
\end{array}
$$

Answer: $42\frac{9}{14}$

The following problem is slightly harder because the two-digit divisor (here 23) is larger.

$$682 \div 23$$

In this problem, the answer is a two-digit number because 682 falls between $23 \times 10 = 230$ and $23 \times 100 = 2300$. To figure out the tens digit of the two-digit answer, you need to ask how many times 23 goes into 680. If you try 30, you'll see that it's slightly too much, as $30 \times 23 = 690$. Now you know that the answer is 20 something, and you can say so. Then subtract $23 \times 20 = 460$ from 682 to obtain 222. Since $23 \times 9 = 207$, the answer is 29 with a remainder of $222 - 207 = 15$.

$$
\begin{array}{r}
29 \\
23\overline{)682} \\
-460 \\
\hline
222 \\
-207 \\
\hline
15
\end{array}
$$

Answer: $29\frac{15}{23}$

Now consider:

$$491 \div 62$$

Since 491 is less than $62 \times 10 = 620$, your answer will simply be a one-digit number with a remainder. You might guess 8, but $62 \times 8 = 496$, which is a little high. Since $62 \times 7 = 434$, the answer is 7 with a remainder of $491 - 434 = 57$, or $7\frac{57}{62}$.

$$
\begin{array}{r}
7 \\
62\overline{)491} \\
-434 \\
\hline
57
\end{array}
$$

$$\text{Answer: } 7\frac{57}{62}$$

Actually, there's a nifty trick to make problems like this easier. Remember how you first tried multiplying 62×8, but found it came out a little high at 496? Well, that wasn't a wasted effort. Aside from knowing that the answer is 7, you can also compute the remainder right away. Since 496 is 5 more than 491, the remainder will be 5 less than 62, the divisor. Since $62 - 5 = 57$, your answer is $7\frac{57}{62}$. The reason this trick works is because $491 = (62 \times 8) - 5 = 62 \times (7 + 1) - 5 = (62 \times 7 + 62) - 5 = (62 \times 7) + (62 - 5) = 62 \times 7 + 57$.

Now try $380 \div 39$ using the shortcut we just learned. So $39 \times 10 = 390$, which is too high by 10. Hence the answer is 9 with a remainder of $39 - 10 = 29$.

Your next challenge is to divide a two-digit number into a four-digit number:

$$3657 \div 54$$

Since $54 \times 100 = 5400$, you know your answer will be a two-digit number. To arrive at the first digit of the answer, you need to figure how many times 54 goes into 3657. Since $54 \times 70 = 3780$ is a little high, you know the answer must be 60 something.

Next, multiply $54 \times 60 = 3240$ and subtract $3657 - 3240 = 417$. Once you say the 60, your problem has been simplified to $417 \div 54$. Since $54 \times 8 = 432$ is a little too high, your last digit is 7 with the remainder $54 - 15 = 39$.

$$67$$
$$54\overline{)3657}$$
$$-3240$$
$$417$$
$$-\ 378$$
$$39$$

Answer: $67\dfrac{39}{54}$

Now try your hand at a problem with a three-digit answer:

$$9467 \div 13$$

$$728$$
$$13\overline{)9467}$$
$$-9100$$
$$367$$
$$-\ 260$$
$$107$$
$$-\ 104$$
$$3$$

Answer: $728\dfrac{3}{13}$

Simplifying Division Problems

If by this point you're suffering from brain strain, relax. As promised, I want to share with you a couple of tricks for simplifying certain mental division problems. These tricks are based on the principle of dividing both parts of the problem by a common factor. If both numbers in the problem are even numbers, you can make the problem twice as easy by dividing each number by 2 before you begin. For example, $858 \div 16$ has two even numbers, and dividing each by 2 yields the much simpler problem of $429 \div 8$:

53	**Divide by 2**	53
16)858		8)429
−800		−400
58		29
− 48		− 24
10		5

$$\textbf{Answer: } 53\frac{10}{16} \qquad\qquad \textbf{Answer: } 53\frac{5}{8}$$

As you can see, the remainders 10 and 5 are not the same; but if you write the remainder in the form of a fraction, you get that $\frac{10}{16}$ is the same as $\frac{5}{8}$. Therefore, when using this method, you must always express the answer in fractional form.

We've done both sets of calculations for you to see how much easier it is. Now you try one for practice:

$$3618 \div 54$$

67	**Divide by 2**	67
54)3618		27)1809
−3240		−1620
378		189
− 378		− 189
0		0

Answer: 67

The problem on the right is much easier to calculate mentally. If you're really alert, you could divide both sides of the original problem by 18 to arrive at an even simpler problem: $201 \div 3 = 67$.

Watch for problems that can be divided by 2 twice, such as $1652 \div 36$:

$$1652 \div 36 \underset{\div 2}{=} 826 \div 18 \underset{\div 2}{=} 413 \div 9 = \begin{array}{r} 45 \\ 9\overline{)413} \\ -360 \\ \hline 53 \\ -\ 45 \\ \hline 8 \end{array}$$

$$\textbf{Answer: } 45\frac{8}{9}$$

I usually find it easier to divide the problem by 2 twice than to divide both numbers by 4. Next, when both numbers end in 0, you can divide each by 10:

$$580 \div 70 \underset{\div 10}{=} 58 \div 7 = \begin{array}{r} 8 \\ 7\overline{)58} \\ -56 \\ \hline 2 \end{array}$$

$$\textbf{Answer: } 8\frac{2}{7}$$

But if both numbers end in 5, double them and then divide both by 10 to simplify the problem. For example:

$$475 \div 35 \underset{\times 2}{=} 950 \div 70 \underset{\div 10}{=} 95 \div 7 = \begin{array}{r} 13 \\ 7\overline{)95} \\ -70 \\ \hline 25 \\ -21 \\ \hline 4 \end{array}$$

$$\text{Answer: } 13\frac{4}{7}$$

Finally, if the divisor ends in 5 and the number you're dividing into ends in 0, multiply both by 2 and then divide by 10, just as you did above:

$$890 \div 45 \underset{\times 2}{=} 1780 \div 90 \underset{\div 10}{=} 178 \div 9 = \begin{array}{r} 19 \\ 9\overline{)178} \\ -90 \\ \hline 88 \\ -81 \\ \hline 7 \end{array}$$

$$\text{Answer: } 19\frac{7}{9}$$

EXERCISE: TWO-DIGIT DIVISION

Here you'll find a variety of two-digit division problems that will test your mental prowess, using the simplification techniques explained earlier in this chapter. Check the end of the book for answers and explanations.

1. $738 \div 17$ 2. $591 \div 24$ 3. $321 \div 79$

4. $4268 \div 28$ 5. $7214 \div 11$ 6. $3074 \div 18$

MATCHING WITS WITH A CALCULATOR: LEARNING DECIMALIZATION

As you might guess, I like to work some magic when I convert fractions to decimals. In the case of one-digit fractions, the best way is to commit the following fractions—from halves through elevenths—to memory. This isn't as hard as it sounds. As you'll see below, most one-digit fractions have special properties that make them hard to forget. Anytime you can reduce a fraction to one you already know, you'll speed up the process.

Chances are you already know the decimal equivalent of the following fractions:

$$\frac{1}{2} = .50 \qquad \frac{1}{3} = .333\ldots \qquad \frac{2}{3} = .666\ldots$$

Likewise:

$$\frac{1}{4} = .25 \qquad \frac{2}{4} = \frac{1}{2} = .50 \qquad \frac{3}{4} = .75$$

The fifths are easy to remember:

$$\frac{1}{5} = .20 \qquad \frac{2}{5} = .40 \qquad \frac{3}{5} = .60 \qquad \frac{4}{5} = .80$$

The sixths require memorizing only two new answers:

$$\frac{1}{6} = .1666\ldots \qquad \frac{2}{6} = \frac{1}{3} = .333\ldots \qquad \frac{3}{6} = \frac{1}{2} = .50$$

$$\frac{4}{6} = \frac{2}{3} = .666\ldots \qquad \frac{5}{6} = .8333\ldots$$

I'll return to the sevenths in a moment. The eighths are a breeze:

$$\frac{1}{8} = .125 \qquad \frac{2}{8} = \frac{1}{4} = .25$$

$$\frac{3}{8} = .375 \ (3 \times \frac{1}{8} = 3 \times .125 = .375) \qquad \frac{4}{8} = \frac{1}{2} = .50$$

$$\frac{5}{8} = .625 \ (5 \times \frac{1}{8} = 5 \times .125 = .625) \qquad \frac{6}{8} = \frac{3}{4} = .75$$

$$\frac{7}{8} = .875 \ (7 \times \frac{1}{8} = 7 \times .125 = .875)$$

The ninths have a magic all their own:

$$\frac{1}{9} = .\overline{1} \qquad \frac{2}{9} = .\overline{2} \qquad \frac{3}{9} = .\overline{3} \qquad \frac{4}{9} = .\overline{4}$$

$$\frac{5}{9} = .\overline{5} \qquad \frac{6}{9} = .\overline{6} \qquad \frac{7}{9} = .7 \qquad \frac{8}{9} = .\overline{8}$$

where the bar indicates that the decimal repeats. For instance, $\frac{4}{9} = .\overline{4} = .444. \ldots$ The tenths you already know:

$$\frac{1}{10} = .10 \qquad \frac{2}{10} = .20 \qquad \frac{3}{10} = .30$$

$$\frac{4}{10} = .40 \qquad \frac{5}{10} = .50 \qquad \frac{6}{10} = .60$$

$$\frac{7}{10} = .70 \qquad \frac{8}{10} = .80 \qquad \frac{9}{10} = .90$$

For the elevenths, if you remember that $\frac{1}{11} = .0909$, the rest is easy:

$$\frac{1}{11} = .\overline{09} = .0909\ldots \qquad \frac{2}{11} = .\overline{18} \ (2 \times .0909)$$

$$\frac{3}{11} = .\overline{27} \ (3 \times .0909) \qquad \frac{4}{11} = .\overline{36} \qquad \frac{5}{11} = .\overline{45}$$

$$\frac{6}{11} = .\overline{54} \qquad \frac{7}{11} = .\overline{63} \qquad \frac{8}{11} = .\overline{72}$$

$$\frac{9}{11} = .\overline{81} \qquad \frac{10}{11} = .\overline{90}$$

The sevenths are truly remarkable. Once you memorize $\frac{1}{7} = .\overline{142857}$, you can get all the other sevenths without having to compute them:

$$\frac{1}{7} = .\overline{142857} \qquad \frac{2}{7} = .\overline{285714} \qquad \frac{3}{7} = .\overline{428571}$$

$$\frac{4}{7} = .\overline{571428} \qquad \frac{5}{7} = .\overline{714285} \qquad \frac{6}{7} = .\overline{857142}$$

Note that the same pattern of numbers repeats itself in each fraction. Only the starting point varies. You can figure out the starting point in a flash by multiplying .14 by the numerator. In the case of $\frac{2}{7}$, 2 × .14 = .28, so use the sequence that begins with the 2, namely $.\overline{285714}$. Likewise with $\frac{3}{7}$, since 3 × .14 = .42, use the sequence that begins with 4, namely $.\overline{428571}$. The rest follow in a similar way.

You will have to calculate fractions higher than $\frac{10}{11}$ as you would any other division problem. However, keep your eyes peeled for ways of simplifying such problems. For example, you can simplify the fraction $\frac{18}{34}$ by dividing both numbers by 2, to reduce it to $\frac{9}{17}$, which is easier to compute.

If the denominator of the fraction is an even number, you can simplify the fraction by reducing it in half, even if the numerator is odd. For example:

$$\frac{9}{14} = \frac{4.5}{7}$$

Dividing the numerator and denominator in half reduces it to a sevenths fraction. Although the sevenths sequence previously shown doesn't provide the decimal for $\frac{4.5}{7}$, once you begin the calculation, the number you memorized will pop up:

$$
\begin{array}{r}
.\overline{6428571} \\
7\overline{)4.5000000} \\
-4.2 \\
\hline
3
\end{array}
$$

As you can see, you needn't work out the entire problem. Once you've reduced it to dividing 3 by 7, you can make a great impression on an audience by rattling off this long string of numbers almost instantly!

When the divisor ends in 5, it almost always pays to double the problem, then divide by 10. For example,

$$
\underset{\times 2}{\frac{29}{45}} = \underset{\div 10}{\frac{58}{90}} = \frac{5.8}{9} = .6\overline{44}
$$

Numbers that end in 25 or 75 should be multiplied by 4 before dividing by 100.

$$
\underset{\times 4}{\frac{31}{25}} = \underset{\div 100}{\frac{124}{100}} = 1.24
$$

$$
\underset{\times 4}{\frac{62}{75}} = \underset{\div 100}{\frac{248}{300}} = \frac{2.48}{3} = .82\overline{66}
$$

You can even put this trick to use in the middle of a problem. If your fraction is $\frac{3}{16}$, look what happens:

$$
\begin{array}{r}
.1 \\
16\overline{)3.000} \\
-16 \\
\hline
14
\end{array}
$$

Once the problem is reduced to $\frac{14}{16}$, you can further reduce it to $\frac{7}{8}$, which you know to be .875. Thus $\frac{3}{16}$ = .1875.

EXERCISE: DECIMALIZATION

To solve the following problems, don't forget to employ the various one-digit fractions you already know as decimals. Wherever appropriate, simplify the fraction before converting it to a decimal.

1. $\dfrac{2}{5}$ 2. $\dfrac{4}{7}$ 3. $\dfrac{3}{8}$ 4. $\dfrac{9}{12}$ 5. $\dfrac{5}{12}$ 6. $\dfrac{6}{11}$

7. $\dfrac{14}{24}$ 8. $\dfrac{13}{27}$ 9. $\dfrac{18}{48}$ 10. $\dfrac{10}{14}$ 11. $\dfrac{6}{32}$ 12. $\dfrac{19}{45}$

TESTING FOR DIVISIBILITY

In the last section, we saw how division problems could be simplified when both numbers were divisible by a common factor. We end this chapter with a brief discussion of how to determine whether one number is a factor of another number. Being able to find the factors of a number helps us simplify division problems and can speed up many multiplication problems. This will also be a very useful tool when we get to advanced multiplication, as you will often be looking for ways to factor a two-, three-, or even a five-digit number in the middle of a multiplication problem. Being able to factor these numbers quickly is very handy. And besides, I think some of the rules are just beautiful.

It's easy to test whether a number is divisible by 2. All you need to do is to check if the last digit is even. If the last digit is 2, 4, 6, 8, or 0, the entire number is divisible by 2.

To test whether a number is divisible by 4, check if the two-digit number at the end is divisible by 4. The number 57,852 is a multiple of 4 because 52 = 13 × 4. The number 69,346 is not a multiple of 4 because 46 is not a multiple of 4. The reason this works is because 4 divides evenly into 100 and thus into any multiple of 100. Thus, since 4 divides evenly into 57,800, and 4 divides into 52, we know that 4 divides evenly into their sum, 57,852.

Likewise, since 8 divides into 1000, to test for divisibility by 8, check the last three digits of the number. For the number 14,918, divide 8 into 918. Since this leaves you with a remainder ($918 \div 8 = 114\frac{6}{8}$), the number is not divisible by 8. You could also have observed this by noticing that 18 (the last two digits of 14,918) is not divisible by 4, and since 14,918 is not divisible by 4, it can't be divisible by 8 either.

When it comes to divisibility by 3, here's a cool rule that's easy to remember: A number is divisible by 3 if and only if the sum of its digits are divisible by 3—no matter how many digits are in the number. To test whether 57,852 is divisible by 3, simply add 5 + 7 + 8 + 5 + 2 = 27. Since 27 is a multiple of 3, we know 57,852 is a multiple of 3. The same amazing rule holds true for divisibility by 9. A number is divisible by 9 if and only if its digits sum to a multiple of 9. Hence, 57,852 is a multiple of 9, whereas 31,416, which sums to 15, is not. The reason this works is based on the fact that the numbers 1, 10, 100, 1,000, 10,000, and so on, are all 1 greater than a multiple of 9.

A number is divisible by 6 if and only if it is even and divisible by 3, so it is easy to test divisibility by 6.

To establish whether a number is divisible by 5 is even easier. Any number, no matter how large, is a multiple of 5 if and only if it ends in 5 or 0.

Establishing divisibility by 11 is almost as easy as determining divisibility by 3 or 9. A number is divisible by 11 if and only if you arrive at either 0 or a multiple of 11 when you alternately subtract and add the digits of a number. For instance, 73,958 is not divisible by 11 since $7 - 3 + 9 - 5 + 8 = 16$. However, the numbers 8,492 and 73,194 are multiples of 11, since $8 - 4 + 9 - 2 = 11$ and $7 - 3 + 1 - 9 + 4 = 0$. The reason this works is based, like the rule for 3s and 9s, on the fact that the numbers 1, 100, 10,000, and 1,000,000 are 1 more than a multiple of 11, whereas the numbers 10, 1,000, 100,000, and so on are 1 less than a multiple of 11.

Testing for divisibility by 7 is a bit trickier. If you add or subtract a number that is a multiple of 7 to the number you are testing, and the resulting number is a multiple of 7, then the test is positive. I always choose to add or subtract a multiple of 7 so that the resulting sum or difference ends in 0. For example, to test the number 5292, I subtract 42 (a multiple of 7) to obtain 5250. Next, I get rid of the 0 at the end (since dividing by ten does not affect divisibility by seven), leaving me with 525. Then I repeat the process by adding 35 (a multiple of 7), which gives me 560. When I delete the 0, I'm left with 56, which I know to be a multiple of 7. Therefore, the original number 5292 is divisible by 7.

This method works not only for 7s, but also for any odd number that doesn't end in 5. For example, to test whether 8792 is divisible by 13, subtract $4 \times 13 = 52$ from 8792, to arrive at 8740. Dropping the 0 results in 874. Then add 2 × 13 = 26 to arrive at 900. Dropping the two 0s leaves you with 9, which is clearly not a multiple of 13. Therefore, 8792 is not a multiple of 13.

EXERCISE: TESTING FOR DIVISIBILITY

In this final set of exercises, be especially careful when you test for divisibility by 7 and 17. The rest should be easy for you.

Divisibility by 2

1. **53,428**   2. **293**   3. **7241**   4. **9846**

Divisibility by 4

5. **3932**   6. **67,348**   7. **358**   8. **57,929**

Divisibility by 8

9. **59,366**   10. **73,488**   11. **248**   12. **6111**

Divisibility by 3

13. **83,671**   14. **94,737**   15. **7359**   16. **3,267,486**

Divisibility by 6

17. **5334**   18. **67,386**   19. **248**   20. **5991**

Divisibility by 9

21. **1234**   22. **8469**   23. **4,425,575**   24. **314,159,265**

Divisibility by 5

25. **47,830**   26. **43,762**   27. **56,785**   28. **37,210**

Divisibility by 11

29. **53,867**   30. **4969**   31. **3828**   32. **941,369**

Divisibility by 7

33. **5784**   34. **7336**   35. **875**   36. **1183**

Divisibility by 17

37. **694**   38. **629**   39. **8273**   40. **13,855**

FRACTIONS

If you can manipulate whole numbers, then doing arithmetic with fractions is almost as easy. In this section, we review the basic methods for adding, subtracting, multiplying, dividing, and simplifying fractions. Those already familiar with fractions can skip this section with no loss of continuity.

Multiplying Fractions

To multiply two fractions, simply multiply the top numbers (called the numerators), then multiply the bottom numbers (called the denominators). For example,

$$\frac{2}{3} \times \frac{4}{5} = \frac{8}{15} \qquad \frac{1}{2} \times \frac{5}{9} = \frac{5}{18}$$

What could be simpler! Try these exercises before going further.

EXERCISE: MULTIPLYING FRACTIONS

1. $\frac{3}{5} \times \frac{2}{7}$ 2. $\frac{4}{9} \times \frac{11}{7}$ 3. $\frac{6}{7} \times \frac{3}{4}$ 4. $\frac{9}{10} \times \frac{7}{8}$

Dividing Fractions

Dividing fractions is just as easy as multiplying fractions. There's just one extra step. First, turn the second fraction upside down (this is called the *reciprocal*) then multiply. For instance, the reciprocal of $\frac{4}{5}$ is $\frac{5}{4}$. Therefore,

$$\frac{2}{3} \div \frac{4}{5} = \frac{2}{3} \times \frac{5}{4} = \frac{10}{12}$$

$$\frac{1}{2} \div \frac{5}{9} = \frac{1}{2} \times \frac{9}{5} = \frac{9}{10}$$

EXERCISE: DIVIDING FRACTIONS

Now it's your turn. Divide these fractions.

1. $\frac{2}{5} \div \frac{1}{2}$
2. $\frac{1}{3} \div \frac{6}{5}$
3. $\frac{2}{5} \div \frac{3}{5}$

Simplifying Fractions

Fractions can be thought of as little division problems. For instance, $\frac{6}{3}$ is the same as $6 \div 3 = 2$. The fraction $\frac{1}{4}$ is the same as $1 \div 4$ (which is .25 in decimal form). Now we know that when we multiply any number by 1, the number stays the same. For example, $\frac{3}{5} = \frac{3}{5} \times 1$. But if we replace 1 with $\frac{2}{2}$, we get $\frac{3}{5} = \frac{3}{5} \times 1 = \frac{3}{5} \times \frac{2}{2} = \frac{6}{10}$. Hence, $\frac{3}{5} = \frac{6}{10}$. Likewise, if we replace 1 with $\frac{3}{3}$, we get $\frac{3}{5} = \frac{3}{5} \times \frac{3}{3} = \frac{9}{15}$. In other words, if we multiply the numerator and denominator by the same number, we get a fraction that is equal to the first fraction.

For another example,

$$\frac{2}{3} = \frac{2}{3} \times \frac{5}{5} = \frac{10}{15}$$

It is also true that if we *divide* the numerator and denominator by the same number, then we get a fraction that is equal to the first one.

For instance,

$$\frac{4}{6} = \frac{4}{6} \div \frac{2}{2} = \frac{2}{3}$$

$$\frac{25}{35} = \frac{25}{35} \div \frac{5}{5} = \frac{5}{7}$$

This is called *simplifying* the fraction.

EXERCISE: SIMPLIFYING FRACTIONS

For the fractions below, can you find an equal fraction whose denominator is 12?

1. $\dfrac{1}{3}$ 2. $\dfrac{5}{6}$ 3. $\dfrac{3}{4}$ 4. $\dfrac{5}{2}$

Simplify these fractions.

5. $\dfrac{8}{10}$ 6. $\dfrac{6}{15}$ 7. $\dfrac{24}{36}$ 8. $\dfrac{20}{36}$

Adding Fractions

The easy case is that of equal denominators. If the denominators are equal, then we add the numerators and keep the same denominators.

For instance,

$$\frac{3}{5} + \frac{1}{5} = \frac{4}{5} \qquad\qquad \frac{4}{7} + \frac{2}{7} = \frac{6}{7}$$

Sometimes we can simplify our answer. For instance,

$$\frac{1}{8} + \frac{5}{8} = \frac{6}{8} = \frac{3}{4}$$

EXERCISE: ADDING FRACTIONS (EQUAL DENOMINATORS)

---

1. $\dfrac{2}{9} + \dfrac{5}{9}$ 2. $\dfrac{5}{12} + \dfrac{4}{12}$ 3. $\dfrac{5}{18} + \dfrac{6}{18}$ 4. $\dfrac{3}{10} + \dfrac{3}{10}$

The trickier case: unequal denominators. When the denominators are not equal, then we replace our fractions with fractions where the denominators are equal.

For instance, to add

$$\frac{1}{3} + \frac{2}{15}$$

we notice that

$$\frac{1}{3} = \frac{5}{15}$$

Therefore,

$$\frac{1}{3} + \frac{2}{15} = \frac{5}{15} + \frac{2}{15} = \frac{7}{15}$$

To add

$$\frac{1}{2} + \frac{7}{8}$$

we notice that

$$\frac{1}{2} = \frac{4}{8}$$

So,

$$\frac{1}{2} + \frac{7}{8} = \frac{4}{8} + \frac{7}{8} = \frac{11}{8}$$

To add

$$\frac{1}{3} + \frac{2}{5}$$

we see that

$$\frac{1}{3} = \frac{5}{15} \text{ and } \frac{2}{5} = \frac{6}{15}$$

So,

$$\frac{1}{3} + \frac{2}{5} = \frac{5}{15} + \frac{6}{15} = \frac{11}{15}$$

EXERCISE: ADDING FRACTIONS (UNEQUAL DENOMINATORS)

1. $\dfrac{1}{5} + \dfrac{1}{10}$ 2. $\dfrac{1}{6} + \dfrac{5}{18}$ 3. $\dfrac{1}{3} + \dfrac{1}{5}$ 4. $\dfrac{2}{7} + \dfrac{5}{21}$

5. $\dfrac{2}{3} + \dfrac{3}{4}$ 6. $\dfrac{3}{7} + \dfrac{3}{5}$ 7. $\dfrac{2}{11} + \dfrac{5}{9}$

Subtracting Fractions

Subtracting fractions works very much like adding them. We have illustrated with examples and provided exercises for you to do.

$$\frac{3}{5} - \frac{1}{5} = \frac{2}{5} \qquad \frac{4}{7} - \frac{2}{7} = \frac{2}{7} \qquad \frac{5}{8} - \frac{1}{8} = \frac{4}{8} = \frac{1}{2}$$

$$\frac{1}{3} - \frac{2}{15} = \frac{5}{15} - \frac{2}{15} = \frac{3}{15} = \frac{1}{5}$$

$$\frac{7}{8} - \frac{1}{2} = \frac{7}{8} - \frac{4}{8} = \frac{3}{8}$$

$$\frac{1}{2} - \frac{7}{8} = \frac{4}{8} - \frac{7}{8} = \frac{-3}{8} \qquad \frac{2}{7} - \frac{1}{4} = \frac{8}{28} - \frac{7}{28} = \frac{1}{28}$$

$$\frac{2}{3} - \frac{5}{8} = \frac{16}{24} - \frac{15}{24} = \frac{1}{24}$$

EXERCISE: SUBTRACTING FRACTIONS

1. $\frac{8}{11} - \frac{3}{11}$　　2. $\frac{12}{7} - \frac{8}{7}$　　3. $\frac{13}{18} - \frac{5}{18}$

4. $\frac{4}{5} - \frac{1}{15}$　　5. $\frac{9}{10} - \frac{3}{5}$　　6. $\frac{3}{4} - \frac{2}{3}$

7. $\frac{7}{8} - \frac{1}{16}$　　8. $\frac{4}{7} - \frac{2}{5}$　　9. $\frac{8}{9} - \frac{1}{2}$

Chapter 5

Good Enough: The Art of "Guesstimation"

So far you've been perfecting the mental techniques necessary to figure out the exact answers to math problems. Often, however, all you'll want is a ballpark estimate. Say you're getting quotes from different lenders on refinancing your home. All you really need at this information-gathering stage is a ballpark estimate of what your monthly payments will be. Or say you're settling a restaurant bill with a group of friends and you don't want to figure each person's bill to the penny. The guesstimation methods described in this chapter will make both these tasks—and many more just like them—much easier. Addition, subtraction, division, and multiplication all lend themselves to guesstimation. As usual, you'll do your computations from left to right.

ADDITION GUESSTIMATION

Guesstimation is a good way to make your life easier when the numbers of a problem are too long to remember. The trick is to round the original numbers up or down:

$$
\begin{array}{r}
8,367 \\
+\ 5,819 \\
\hline
14,186
\end{array}
\quad \approx \quad
\begin{array}{r}
8,000 \\
+\ 6,000 \\
\hline
14,000
\end{array}
$$

(≈ means approximately)

George Parker Bidder: The Calculating Engineer

The British have had their share of lightning calculators, and the mental performances of George Parker Bidder (1806–1878), born in Devonshire, were as impressive as any. Like most lightning calculators, Bidder began to try his hand (and mind) at mental arithmetic as a young lad. Learning to count, add, subtract, multiply, and divide by playing with marbles, Bidder went on tour with his father at age nine.

Almost no question was too difficult for him to handle. "If the moon is 123,256 miles from the earth and sound travels four miles a minute, how long would it take for sound to travel from the earth to the moon?" The young Bidder, his face wrinkled in thought for nearly a minute, replied, "Twenty-one days, nine hours, thirty-four minutes." (We know now that the distance is closer to 240,000 miles and sound cannot travel through the vacuum of space.) At age ten, Bidder mentally computed the square root of 119,550,669,121 as 345,761 in a mere thirty seconds. In 1818, Bidder and the American lightning calculator Zerah Colburn were paired in a mental calculating duel in which Bidder, apparently, "outnumbered" Colburn.

Riding on his fame, George Bidder entered the University of Edinburgh and went on to become one of the more respected engineers in England. In parliamentary debates over railroad conflicts, Bidder was frequently called as a witness, which made the opposition shudder; as one said, "Nature had endowed him with particular qualities that did not place his opponents on a fair footing." Unlike Colburn, who retired as a lightning calculator at age twenty, Bidder kept it up for his entire life. As late as 1878, in fact, just before his death, Bidder calculated the number of vibrations of light striking the eye in one second, based on the fact that there are 36,918 waves of red light per inch, and light travels at approximately 190,000 miles per second.

Notice that we rounded the first number down to the nearest thousand and the second number up. Since the exact answer is 14,186, our relative error is small.

If you want to be more exact, instead of rounding off to the nearest thousand, round off to the nearest hundred:

$$
\begin{array}{r}
8,367 \\
+\ 5,819 \\
\hline
14,186
\end{array}
\quad\approx\quad
\begin{array}{r}
8,400 \\
+\ 5,800 \\
\hline
14,200
\end{array}
$$

The answer is only 14 off from the exact answer, an error of less than .1%. This is what I call a good guesstimation!

Try a five-digit addition problem, rounding to the nearest hundred:

$$
\begin{array}{r}
46,187 \\
+\ 19,378 \\
\hline
65,565
\end{array}
\quad\approx\quad
\begin{array}{r}
46,200 \\
+\ 19,400 \\
\hline
65,600
\end{array}
$$

By rounding to the nearest hundred, our answer will always be off by less than 100. If the answer is larger than 10,000, your guesstimate will be within 1% of the exact answer.

Now let's try something wild:

$$
\begin{array}{r}
23,859,379 \\
+\ 7,426,087 \\
\hline
31,285,466
\end{array}
\approx
\begin{array}{r}
24,000,000 \\
+\ 7,000,000 \\
\hline
31,000,000
\end{array}
\ or\
\begin{array}{r}
23.9\ \text{million} \\
+\ 7.4\ \text{million} \\
\hline
31.3\ \text{million}
\end{array}
$$

If you round to the nearest million, you get an answer of 31 million, off by roughly 285,000. Not bad, but you can do better by rounding to the nearest hundred thousand, as we've shown in the right-hand column. Again your guesstimate will be within

1% of the precise answer. If you can compute these smaller problems exactly, you can guesstimate the answer to any addition problem.

Guesstimating at the Supermarket

Let's try a real-world example. Have you ever gone to the store and wondered what the total is going to be before the cashier rings it up? For estimating the total, my technique is to round the prices to the nearest 50¢. For example, while the cashier is adding the numbers shown below on the left, I mentally add the numbers shown on the right:

$ 1.39	$ 1.50
0.87	1.00
2.46	2.50
0.61	0.50
3.29	3.50
2.99	3.00
0.20	0.00
1.17	1.00
0.65	0.50
2.93	3.00
3.19	3.00
$19.75	$19.50

My final figure is usually within a dollar of the exact total.

SUBTRACTION GUESSTIMATION

The way to guesstimate the answers to subtraction problems is the same—you round to the nearest thousand or hundreds digit, preferably the latter:

$$
\begin{array}{ccccc}
8{,}367 & & 8{,}000 & & 8{,}400 \\
-\ 5{,}819 & \approx & -\ 6{,}000 & or & -\ 5{,}800 \\
\hline
2{,}548 & & 2{,}000 & & 2{,}600
\end{array}
$$

You can see that rounding to the nearest thousand leaves you with an answer quite a bit off the mark. By rounding to the second digit (hundreds, in the example), your answer will usually be within 3% of the exact answer. For this problem, your answer is off by only 52, a relative error of 2%. If you round to the third digit, the relative error will usually be below 1%. For instance:

$$
\begin{array}{ccccc}
439{,}412 & & 440{,}000 & & 439{,}000 \\
-\ 24{,}926 & \approx & -\ 20{,}000 & or & -\ 25{,}000 \\
\hline
414{,}486 & & 420{,}000 & & 414{,}000
\end{array}
$$

By rounding the numbers to the third digit rather than to the second digit, you improve the accuracy of the estimate by a significant amount.

DIVISION GUESSTIMATION

The first, and most important, step in guesstimating the answer to a division problem is to determine the magnitude of the answer:

$$
\begin{array}{r}
9{,}644.5 \\
6{\overline{\smash{)}}57{,}867.0}
\end{array}
\quad \approx \quad
\begin{array}{r}
9 \\
6{\overline{\smash{)}}58{,}000} \\
\underline{54} \\
4
\end{array}
$$

$$\text{Answer} \approx 9\frac{2}{3} \text{ thousand} = 9{,}667$$

The next step is to round off the larger numbers to the nearest thousand and change the 57,867 to 58,000. Dividing 6 into 58 gives you 9 with a remainder. But the most important component in this problem is where to place the 9.

For example, multiplying 6 × 90 yields 540, while multiplying 6 × 900 yields 5,400, both of which are too small. But 6 × 9,000 = 54,000, which is pretty close to the answer. This tells you the answer is 9,000 and something. You can estimate just what that something is by first subtracting 58 − 54 = 4. At this point you could bring down the 0 and divide 6 into 40, and so forth. But if you're on your toes, you'll realize that dividing 6 into 4 gives you $\frac{4}{6} = \frac{2}{3} \approx .667$. Since you know the answer is 9,000 something, you're now in a position to guess 9,667. In fact, the actual answer is 9,645—darn close!

Division on this level is simple. But what about large division problems? Let's say we want to compute, just for fun, the amount of money a professional athlete earns a day if he makes $5,000,000 a year:

$$\text{365 days}\overline{)\$5,000,000}$$

First you must determine the magnitude of the answer. Does this player earn thousands every day? Well, 365 × 1000 = 365,000, which is too low.

Does he earn tens of thousands every day? Well, 365 × 10,000 = 3,650,000, and that's more like it. To guesstimate your answer, divide the first two digits (or 36 into 50) and figure that's $1\frac{14}{36}$, or $1\frac{7}{18}$. Since 18 goes into 70 about 4 times, your guess is that the athlete earns about $14,000. The exact answer is $13,698.63 per day. Not a bad estimate (and not a bad salary!).

Here's an *astronomical* calculation for you. How many seconds does it take light to get from the sun to the earth? Well, light travels at 186,282 miles per second, and the sun is (on average) 92,960,130 miles away. I doubt you're particularly eager to attempt this problem by hand. Fortunately, it's relatively simple to guesstimate an answer. First, simplify the problem:

$$186{,}282\overline{)92{,}960{,}130} \approx 186\overline{)93{,}000}$$

Now divide 186 into 930, which yields 5 with no remainder. Then append the two 0s you removed from 93,000 and you get 500 seconds. The exact answer is 499.02 seconds, so this is a very respectable guesstimate.

MULTIPLICATION GUESSTIMATION

You can use much the same techniques to guesstimate your answers to multiplication problems. For example,

$$
\begin{array}{ccc}
88 & & 90 \\
\times\,54 & \approx & \times\,50 \\
\hline
4752 & & 4500 \\
\end{array}
$$

Rounding up to the nearest multiple of 10 simplifies the problem considerably, but you're still off by 252, or about 5%. You can do better if you round both numbers by the same amount, but in opposite directions. That is, if you round 88 by *increasing* 2, you should also *decrease* 54 by 2:

$$
\begin{array}{ccc}
88 & & 90 \\
\times\,54 & \approx & \times\,52 \\
\hline
4752 & & 4680 \\
\end{array}
$$

Instead of a 1-by-1 multiplication problem, you now have a 2-by-1 problem, which should be easy enough for you to do. Your guesstimation is off by only 1.5%.

When you guesstimate the answer to multiplication problems by rounding the larger number up and the smaller number down, your guesstimate will be a little low. If you round the larger number down and the smaller number up so that the numbers are closer together, your guesstimate will be a little high. The larger the amount by which you round up or down, the greater your guesstimate will be off from the exact answer. For example:

$$
\begin{array}{ccc}
73 & & 70 \\
\times\,65 & \approx & \times\,68 \\
\hline
4745 & & 4760
\end{array}
$$

Since the numbers are closer together after you round them off, your guesstimate is a little high.

$$
\begin{array}{ccc}
67 & & 70 \\
\times\,67 & \approx & \times\,64 \\
\hline
4489 & & 4480
\end{array}
$$

Since the numbers are farther apart, the estimated answer is too low, though again, not by much. You can see that this multiplication guesstimation method works quite well. Also notice that this problem is just 67^2 and that our approximation is just the first step of the squaring techniques. Let's look at one more example:

$$
\begin{array}{ccc}
83 & & 85 \\
\times\,52 & \approx & \times\,50 \\
\hline
4316 & & 4250
\end{array}
$$

We observe that the approximation is most accurate when the original numbers are close together. Try estimating a 3-by-2 multiplication problem:

$$
\begin{array}{r}
728 \\
\times\ 63 \\
\hline
45{,}864
\end{array}
\quad \approx \quad
\begin{array}{r}
731 \\
\times\ 60 \\
\hline
43{,}860
\end{array}
$$

By rounding 63 down to 60 and 728 up to 731, you create a 3-by-1 multiplication problem, which puts your guesstimate within 2004 of the exact answer, an error of 4.3%.

Now try guesstimating the following 3-by-3 problem:

$$
\begin{array}{r}
367 \\
\times\ 492 \\
\hline
180{,}564
\end{array}
\quad \approx \quad
\begin{array}{r}
359 \\
\times\ 500 \\
\hline
179{,}500
\end{array}
$$

You will notice that although you rounded both numbers up and down by 8, your guesstimate is off by over 1000. That's because the multiplication problem is larger and the size of the rounding number is larger, so the resulting estimate will be off by a greater amount. But the relative error is still under 1%.

How high can you go with this system of guesstimating multiplication problems? As high as you want. You just need to know the names of large numbers. A thousand thousand is a million, and a thousand million is a billion. Knowing these names and numbers, try this one on for size:

$$
\begin{array}{r}
28{,}657{,}493 \\
\times\ 13{,}864 \\
\hline
\end{array}
\quad \approx \quad
\begin{array}{r}
29\ \text{million} \\
\times\ 14\ \text{thousand} \\
\hline
\end{array}
$$

As before, the objective is to round the numbers to simpler numbers such as 29,000,000 and 14,000. Dropping the 0s for now, this is just a 2-by-2 multiplication problem: $29 \times 14 = 406$ $(29 \times 14 = 29 \times 7 \times 2 = 203 \times 2 = 406)$. Hence the answer is roughly 406 billion, since a thousand million is a billion.

SQUARE ROOT ESTIMATION: DIVIDE AND AVERAGE

The quantity $\sqrt{n}$, the square root of a number n, is the number which, when multiplied by itself, will give you n. For example, the square root of 9 is 3 because $3 \times 3 = 9$. The square root is used in many science and engineering problems and is almost always solved with a calculator. The following method provides an accurate estimate of the answer.

In square root estimation your goal is to come up with a number that when multiplied by itself approximates the original number. Since the square root of most numbers is not a whole number, your estimate is likely to contain a fraction or decimal point.

Let's start by guesstimating the square root of 19. Your first step is to think of the number that when multiplied by itself comes closest to 19. Well, $4 \times 4 = 16$ and $5 \times 5 = 25$. Since 25 is too high, the answer must be 4 point something. Your next step is to divide 4 into 19, giving you 4.75. Now, since 4×4 is less than $4 \times 4.75 = 19$, which in turn is less than 4.75×4.75, we know that 19 (or 4×4.75) lies between 4^2 and 4.75^2. Hence, the square root of 19 lies between 4 and 4.75.

I'd guess the square root of 19 to be about halfway between, at 4.375. In fact, the square root of 19 (rounded to three decimal places) is 4.359, so our guesstimate is pretty close. We illustrate this procedure as follows:

Divide:	Average:

$$\begin{array}{r} 4.75 \\ 4\overline{)19.0} \\ \underline{16} \\ 30 \\ \underline{28} \\ 20 \\ \underline{20} \\ 0 \end{array}$$

$$\frac{4 + 4.75}{2} = 4.375$$

Actually, we can obtain this answer another way, which you might find easier. We know 4 squared is 16, which is shy of 19 by 3. To improve our guess, we *"add the error divided by twice our guess."* Here, we add 3 divided by 8 to get $4\frac{3}{8} = 4.375$. We note that this method will always produce an answer that is a little higher than the exact answer.

Now you try a slightly harder one. What's the square root of 87?

Divide:	Average:

$$\begin{array}{r} 9.\overline{66} \\ 9\overline{)87.0} \end{array}$$

$$\frac{9 + 9.66}{2} = 9.\overline{33}$$

First come up with your ballpark figure, which you can get fairly quickly by noting that $9 \times 9 = 81$ and $10 \times 10 = 100$, which means the answer is 9 point something. Carrying out the division of 9 into 87 to two decimal places, you get 9.66. To improve your guesstimate, take the average of 9 and 9.66, which is 9.33—exactly the square root of 87 rounded to the second decimal place! Alternatively, our guesstimate is $9 + (\text{error})/18 = 9 + \frac{6}{18} = 9.\overline{33}$.

Using this technique, it's pretty easy to guesstimate the square root of two-digit numbers. But what about three-digit numbers? Actually, they are not much harder. I can tell you right off the bat that all three-digit and four-digit numbers have two-digit square roots before the decimal point. And the procedure for computing square roots is the same, no matter how large the number. For instance, to compute the square root of 679, first find your ballpark figure. Because 20 squared is 400 and 30 squared is 900, the square root of 679 must lie between 20 and 30.

When you divide 20 into 679, you get approximately 34. Averaging 20 and 34 gives you a guesstimate of 27, but here's a better estimate. If you know that 25 squared is 625, then your error is $679 - 625 = 54$. Dividing that by 50, we have $\frac{54}{50} = \frac{108}{100} = 1.08$. Hence our improved guesstimate is $25 + 1.08 = 26.08$. (For an even better estimate, if you know that 26 squared is 676, your error is 3, so add $\frac{3}{52} \approx .06$ to get 26.06.) The exact answer is 26.06, rounded to two decimal places.

To guesstimate the square root of four-digit numbers, look at the first two digits of the number to determine the first digit of the square root. For example, to find the square root of 7369, consider the square root of 73. Since $8 \times 8 = 64$ and $9 \times 9 = 81$, 8 must be the first digit of the square root. So the answer is 80 something. Now proceed the usual way. Dividing 80 into 7369 gives 92 plus a fraction, so a good guesstimate is 86. If you squared 86 to get 7396, you would be high by 27, so you should subtract $\frac{27}{172} \approx .16$ for a better guesstimate of 85.84, which is right on the money.

To guesstimate the square root of a six-digit number like 593,472 would seem like an impossible task for the uninitiated, but for you it's no sweat. Since $700^2 = 490,000$, and $800^2 = 640,000$, the square of 593,472 must lie between 700 and 800. In fact, all five-digit and six-digit numbers have three-digit square roots. In practice, you only need to look at the square root of the first two digits of six-digit numbers (or the first digit of five numbers). Once you figure out that the square root of 59 lies between 7 and 8, you know your answer is in the 700s.

Now proceed in the usual manner:

Divide:

$$\begin{array}{r} 847 \\ 700\overline{)593472} \end{array} \approx \begin{array}{r} 847 \\ 7\overline{)5934} \end{array}$$

Average:

$$\frac{700 + 847}{2} = 773.5$$

The exact square root of 593,472 is 770.37 (to five places), so you're pretty close. But you could have been closer, as the following trick demonstrates. Note that the first two digits, 59, are closer to 64 (8 × 8) than they are to 49 (7 × 7). Because of this you can start your guesstimation with the number 8 and proceed from there:

Divide:

$$\begin{array}{r} 741 \\ 800\overline{)593472} \end{array} \approx \begin{array}{r} 741 \\ 8\overline{)5934} \end{array}$$

Average:

$$\frac{800 + 741}{2} = 770.5$$

Just for fun, let's do a real whopper—the square root of 28,674,529. This isn't as hard as it might seem. Your first step is to round to the nearest large number—in this case, just find the square root of 29.

Divide:

$$\begin{array}{r} 5.8 \\ 5\overline{)29.0} \\ \underline{25} \\ 40 \end{array}$$

Average:

$$\frac{5 + 5.8}{2} = 5.4$$

All seven-digit and eight-digit numbers have four-digit square roots, so 5.4 becomes 5400, your estimate. The exact answer is slightly greater than 5354.8. Not bad!

This wraps up the chapter on guesstimation math. After doing the exercises below, turn to the next chapter on pencil-and-paper math, where you will learn to write down answers to problems, but in a much quicker way than you've done on paper before.

The Mathematical Duel of Évariste Galois

The tragic story of the French mathematician Évariste Galois (1811–1832) killed at the age of twenty in a duel over "an infamous coquette" is legendary in the annals of the history of mathematics. A precociously brilliant student, Galois laid the foundation for a branch of mathematics known as group theory. Legend has it that he penned his theory the night before the duel, anticipating his demise and wanting to leave his legacy to the mathematics community. Hours before his death, on May 30, 1832, Galois wrote to Auguste Chevalier: "I have made some new discoveries in analysis. The first concerns the theory of equations, the others integral functions." After describing these, he asked his friend: "Make a public request of Jacobi or Gauss to give their opinions not as to the truth but as to the importance of these theorems. After that, I hope some men will find it profitable to sort out this mess."

Romantic legend and historical truth, however, do not always match. What Galois penned the night before his death were corrections and editorial changes to papers that had been accepted by the Academy of Sciences long before. Further, Galois's initial papers had been submitted three years prior to the duel, when he was all of seventeen! It was after this that Galois became embroiled in political controversy, was arrested, spent time in a prison dungeon, and, ultimately, got himself mixed up in a dispute over a woman and killed.

Aware of his own precocity, Galois noted, "I have carried out researches which will halt many savants in theirs." For over a century that proved to be the case.

MORE TIPS ON TIPS

As we indicated in Chapter 0, it is easy to figure out tips for most situations. For example, to figure out a 10% tip, we merely multiply the bill by 0.1 (or divide the bill by 10). For example, if a bill came to $42, then a 10% tip would be $4.20. To determine a 20% tip, you simply multiply the bill by 0.2, or double the amount of a 10% tip. Thus a 20% tip on a $42 bill would be $8.40.

To determine a 15% tip, we have a few options. If you have mastered the techniques of Chapter 2, and are comfortable with multiplying by $15 = 5 \times 3$, you can simply multiply the bill by 15, then divide by 100. For example, with a $42 bill, we have $42 \times 15 = 42 \times 5 \times 3 = 210 \times 3 = 630$, which when divided by 100, gives us a tip of $6.30. Another method is to take the average of a 10% tip and a 20% tip. From our earlier calculation, this would be

$$\frac{\$4.20 + \$8.40}{2} = \frac{\$12.60}{2} = \$6.30$$

Perhaps the most popular approach to taking a 15% tip is to take 10% of the bill, cut that amount in half (which would be a 5% tip), then add those two numbers together. So, for instance, with a $42 bill, you would add $4.20 plus half that amount, $2.10, to get

$$\$4.20 + \$2.10 = \$6.30$$

Let's use all three methods to compute 15% on a bill that is $67. By the direct method, $67 \times 3 \times 5 = 201 \times 5 = 1005$, which when divided by 100 gives us $10.05. By the averaging method, we average the 10% tip of $6.70 with the 20% tip of $13.40, to get

$$\frac{\$6.70 + \$13.40}{2} = \frac{\$20.10}{2} = \$10.05$$

Using the last method, we add $6.70 to half that amount, $3.35 to get

$$\$6.70 + \$3.35 = \$10.05$$

Finally, to calculate a 25% tip, we offer two methods. Either multiply the amount by 25, then divide by 100, or divide the amount by 4 (perhaps by cutting the amount in half twice). For example, with a $42 bill, you can compute $42 \times 25 = 42 \times 5 \times 5 = 210 \times 5 = 1050$, which when divided by 100 produces a tip of $10.50. Or you can divide the original amount directly by 4, or cut in half twice: half of $42 is $21, and half of that is $10.50. With a $67 bill, I would probably divide by 4 directly: since $67 = 4 = 16\frac{3}{4}$, we get a 25% tip of $16.75.

NOT-TOO-TAXING CALCULATIONS

In this section, I will show you my method for mentally estimating sales tax. For some tax rates, like 5% or 6% or 10%, the calculation is straightforward. For example, to calculate 6% tax, just multiply by 6, then divide by 100. For instance, if the sale came to $58, then $58 \times 6 = 348$, which when divided by 100 gives the exact sales tax of $3.48. (So your total amount would be $61.48.)

But how would you calculate a sales tax of $6\frac{1}{2}\%$ on $58?

I will show you several ways to do this calculation, and you choose the one that seems easiest for you. Perhaps the easiest way to add half of a percent to any dollar amount is to simply cut the dollar amount in half, then turn it into cents. For example, with $58, since half of 58 is 29, simply add 29 cents to the 6% tax (already computed as $3.48) to get a sales tax of $3.77.

Another way to calculate the answer (or a good mental estimate) is to take the 6% tax, divide it by 12, then add those two numbers. For example, since 6% of $58 is $3.48, and 12 goes into 348 almost 30 times, then add 30 cents for an estimate of $3.78, which is only off by a penny. If you would rather divide by 10 instead of 12, go ahead and do it. You will be calculating 6.6% instead of 6.5% (since $\frac{6}{10} = 0.6$) but that should still be a very good estimate. Here, you would take $3.48 and add 34¢ to get $3.82.

Let's try some other sales tax percentages. How can we calculate $7\frac{1}{4}$% sales tax on $124? Begin by computing 7% of 124. From the methods shown in Chapter 2, you know that 124 × 7 = 868, so 7% of 124 is $8.68. To add a quarter percent, you can divide the original dollar amount by 4 (or cut it in half, twice) and turn the dollars into cents. Here, 124 = 4 = 31, so add 31¢ to $8.68 to get an exact sales tax of $8.99.

Another way to arrive at 31¢ is to take your 7% sales tax, $8.68, and divide it by 28. The reason this works is because $\frac{7}{28} = \frac{1}{4}$. For a quick mental estimate, I would probably divide $8.68 by 30, to get about 29¢, for an approximate total sales tax of $8.97.

When you divide by 30, then you are actually computing a tax of $7\frac{7}{30}$%, which is approximately 7.23% instead of 7.25%.

How would you calculate a sales tax of 7.75%? Probably for most approximations, it is sufficient to just say that it is a little less than 8% sales tax. But to get a better approximation, here are some suggestions. As you saw in the last example, if you can easily calculate a $\frac{1}{4}$% adjustment, then simply triple that amount for the $\frac{3}{4}$% adjustment. For example, to calculate 7.75% of $124, you first calculate 7% to obtain $8.68. If you calculated that $\frac{1}{4}$% is 31¢, then $\frac{3}{4}$% would be 93¢, for a grand total of $8.68 + 0.93 = $9.61. For a quick approximation, you can exploit the fact that $\frac{7}{9}$ = .777 is approximately .75, so you can divide the 7% sales tax by 9 to get slightly more than the .75% adjustment. In this example, since $8.68 divided by 9 is about 96¢, simply add $8.68 + 0.96 = $9.64 for a slight overestimate.

We can use this approximation procedure for any sales tax. For a general formula, to estimate the sales tax of $A.B%, first multiply the amount of the sale by A%. Then divide this amount by the number D, where A/D is equal to 0.B. (Thus D is equal to A times the reciprocal of B.) Adding these numbers together gives you the total sales tax (or an approximate one, if you rounded D to an easier nearby number). For instance, with 7.75%, the magic divisor D would be $7 \times \frac{4}{3} = \frac{28}{3} = 9\frac{1}{3}$, which we rounded down to 9. For sales tax of $6\frac{3}{8}$%, first compute the sales tax on 6%, then divide that number by 16, since $\frac{6}{16} = \frac{3}{8}$. (To divide a number by 16, divide the number by 4 twice, or divide the number by 8, then by 2.) Try to come up with methods for finding the sales tax in the state that you live in. You will find that the problem is not as taxing as it seems!

SOME "INTEREST-ING" CALCULATIONS

Finally, we'll briefly mention some practical problems pertaining to interest, from the standpoint of watching your investments grow, and paying off money that you owe.

We begin with the famous Rule of 70, which tells you approximately how long it takes your money to double: **To find the**

**number of years that it will take for your money to double,
divide the number 70 by the rate of interest.**

Suppose that you find an investment that promises to pay you
5% interest per year. Since $70 \div 5 = 14$, then it will take about
14 years for your money to double. For example, if you invested
$1000 in a savings account that paid that interest, then after
14 years, it will have $1000(1.05)^{14} = \$1979.93$. With an inter-
est rate of 7%, the Rule of 70 indicates that it will take about
10 years for your money to double. Indeed, if you invest
$1000 at that annual interest rate, you will have after ten years
$1000(1.07)^{10} = \$1967.15$. At a rate of 2%, the Rule of 70 says
that it should take about 35 years to double, as shown below:

$$\$1000(1.02)^{35} = \$1999.88$$

A similar method is called the **Rule of 110**, which indicates
how long it takes for your money to *triple*. For example, at 5%
interest, since $110 \div 5 = 22$, it should take about 22 years to
turn $1000 into $3000. This is verified by the calculation
$1000(1.05)^{22} = \$2925.26$. The Rule of 70 and the Rule of 110
are based on properties of the number $e = 2.71828 \ldots$ and
"natural logarithms" (studied in precalculus), but fortunately we
do not need to utilize this higher mathematics to apply the rules.

Now suppose that you borrow money, and you have to pay the
money back. For example, suppose that you borrow $360,000
at an annual interest rate of 6% per year (which we shall inter-
pret to mean that interest will accumulate at a rate of 0.5% per
month) and suppose that you have thirty years to pay off the
loan. About how much will you need to pay each month? First
of all, you will need to pay $360,000 times 0.5% = $1,800 each
month just to cover what you owe in interest. (Although, actually, what you owe in interest will go down gradually over time.)

Since you will make $30 \times 12 = 360$ monthly payments, then paying an extra $1,000 each month would cover the rest of the loan, so an upper bound on your monthly payment is $1,800 + $1,000 = $2,800. But fortunately you do not need to pay that much extra. Here is my rule of thumb for estimating your monthly payment.

Let i be your *monthly* interest rate. (This is your annual interest rate divided by 12.) Then to pay back a loan of $P in N months, your monthly payment M is about

$$M = \frac{Pi(1 + i)^N}{(1 + i)^N - 1}$$

In our last example, $P = \$360,000$, and $i = 0.005$, so our formula indicates that our monthly payment should be about

$$M = \frac{\$360,000(.005)(1.005)^{360}}{(1.005)^{360} - 1}$$

Notice that the first two numbers in the numerator multiply to $1,800. Using a calculator (for a change) to compute $(1.005)^{360} = 6.02$, we have that your monthly payment should be about $1,800(6.02)/5.02, which is about $2,160 per month.

Here's one more example. Suppose you wish to pay for a car, and after your down payment, you owe $18,000 to be paid off in five years, with an annual interest rate of 4%. If there was no interest, you would have to pay $18,000/60 = $300 per month. Since the first year's interest is $18,000(.04) = $720, you know that you will need to pay no more than $300 + $60 = $360. Using our formula, since the monthly interest rate is $i = .04/12 = 0.00333$, we have

$$M = \frac{\$18,000(0.0333)(1.00333)^{60}}{(1.00333)^{50} - 1}$$

and since $(1.00333)^{60} = 1.22$, we have a monthly payment of about $\$60(1.22)/(0.22) = \$333$.

We conclude with some exercises that we hope will hold your interest.

GUESSTIMATION EXERCISES

Go through the following exercises for guesstimation math; then check your answers and computations at the back of the book.

EXERCISE: ADDITION GUESSTIMATION

Round these numbers up or down and see how close you come to the exact answer:

1.	2.	3.	4.
1479	57,293	312,025	8,971,011
+ 1105	+ 37,421	+ 79,419	+ 4,016,367

Mentally estimate the total for the following column of prices by rounding to the nearest 50¢:

$$
\begin{array}{r}
\$\ 2.67 \\
1.95 \\
7.35 \\
9.21 \\
0.49 \\
11.21 \\
0.12 \\
6.14 \\
\underline{8.31}
\end{array}
$$

EXERCISE: SUBTRACTION GUESSTIMATION

Estimate the following subtraction problems by rounding to the second or third digit:

1.	2.	3.	4.
4,926	**67,221**	**526,978**	**8,349,241**
− 1,659	**− 9,874**	**− 42,009**	**− 6,103,839**

EXERCISE: DIVISION GUESSTIMATION

Adjust the numbers in a way that allows you to guesstimate the following division problems:

1. $7\overline{)4379}$ 2. $5\overline{)23,958}$ 3. $13\overline{)549,213}$

4. $289\overline{)5,102,357}$ 5. $203,637\overline{)8,329,483}$

EXERCISE: MULTIPLICATION GUESSTIMATION

Adjust the numbers in a way that allows you to guesstimate the following multiplication problems:

1.	2.	3.	4.	5.
98	**76**	**88**	**539**	**312**
× 27	**× 42**	**× 88**	**× 17**	**× 98**

6.	7.	8.	9.	10.
639	**428**	**51,276**	**104,972**	**5,462,741**
× 107	**× 313**	**× 489**	**× 11,201**	**× 203,413**

EXERCISE: SQUARE ROOT GUESSTIMATION

Estimate the square roots of the following numbers using the divide and average method:

1. $\sqrt{17}$ 2. $\sqrt{35}$ 3. $\sqrt{163}$ 4. $\sqrt{4279}$ 5. $\sqrt{8039}$

EXERCISE: EVERYDAY MATH

1. Compute 15% of $88.

2. Compute 15% of $53.

3. Compute 25% of $74.

4. How long does it take an annual interest rate of 10% to double your money?

5. How long does it take an annual interest rate of 6% to double your money?

6. How long does it take an annual interest rate of 7% to triple your money?

7. How long does it take an annual interest rate of 7% to quadruple your money?

8. Estimate the monthly payment to repay a loan of $100,000 at an interest rate of 9% over a ten-year period.

9. Estimate the monthly payment to repay a loan of $30,000 at an interest rate of 5% over a four-year period.

Chapter 6

Math for the Board:
Pencil-and-Paper Math

In the Introduction to this book I discussed the many benefits you will get from being able to do mental calculations. In this chapter I present some methods for speeding up pencil-and-paper calculations as well. Since calculators have replaced much of the need for pencil-and-paper arithmetic in most practical situations, I've chosen to concentrate on the lost art of calculating square roots and the flashy criss-cross method for multiplying large numbers. Since these are, admittedly, mostly for mental gymnastics and not for some practical application, I will first touch on addition and subtraction and show you just a couple of little tricks for speeding up the process and for checking your answers. These techniques *can* be used in daily life, as you'll see.

If you are eager to get to the more challenging multiplication problems, you can skip this chapter and go directly to Chapter 7, which is critical for mastering the big problems in Chapter 8. If you need a break and just want to have some fun, then I recommend going through this chapter—you'll enjoy playing with pencil and paper once again.

COLUMNS OF NUMBERS

Adding long columns of numbers is just the sort of problem you might run into in business or while figuring out your personal finances. Add the following column of numbers as you normally would and then check to see how I do it.

$$
\begin{array}{r}
4328 \\
884 \\
620 \\
1477 \\
617 \\
+\ 725 \\
\hline
8651
\end{array}
$$

When I have pencil and paper at my disposal, I add the numbers from top to bottom and from right to left, just as we learned to do in school. With practice, you can do these problems in your head as fast or faster than you can with a calculator. As I sum the digits, the only numbers that I "hear" are the partial sums. That is, when I sum the first (rightmost) column $8 + 4 + 0 + 7 + 7 + 5$, I hear $8 \ldots 12 \ldots 19 \ldots 26 \ldots 31$. Then I put down the 1, carry the 3, and proceed as usual. The next column would then sound like $3 \ldots 5 \ldots 13 \ldots 15 \ldots 22 \ldots 23 \ldots 25$. Once I have my final answer, I write it down, then check my computation by adding the numbers from bottom to top and, I hope, arrive at the same answer.

For instance, the first column would be summed, from bottom to top, as $5 + 7 + 7 + 0 + 4 + 8$ (which in my mind sounds like $5 \ldots 12 \ldots 19 \ldots 23 \ldots 31$). Then I mentally carry the 3, and add $3 + 2 + 1 + 7 + 2 + 8 + 2$, and so on. By adding the numbers in a different order, you are less likely to make the same mistake twice. Of course, if the answers differ, then at least one of the calculations must be wrong.

MOD SUMS

If I'm not sure about my answer, I sometimes check my solution by a method I call mod sums (because it is based on the elegant mathematics of modular arithmetic). This method also goes by the names of digital roots and casting out nines. I admit this method is not as practical, but it's easy to use.

With the mod sums method, you sum the digits of each number until you are left with a single digit. For example, to compute the mod sum of 4328, add $4 + 3 + 2 + 8 = 17$. Then add the digits of 17 to get $1 + 7 = 8$. Hence the mod sum of 4328 is 8. For the previous problem the mod sums of each number are computed as follows:

4328	→ 17	→	8
884	→ 20	→	2
620	→ 8	→	8
1477	→ 19 → 10	→	1
617	→ 14	→	5
+ 725	→ 14	→	+ 5
8651			29

8651 → 20 → 2

29 → 11 → 2

As illustrated above, the next step is to add all the mod sums together (8 + 2 + 8 + 1 + 5 + 5). This yields 29, which sums to 11, which in turn sums to 2. Note that the mod sum of 8651, your original total of the original digits, is also 2. This is not a coincidence! If you have computed the answer and the mod sums correctly, your final mod sums must be the same. If they are different, you have definitely made a mistake somewhere: there is about a 1 in 9 chance that the mod sums will match accidentally. If there is a mistake, then this method will detect it 8 times out of 9.

The mod sum method is more commonly known to mathematicians and accountants as casting out nines because the mod sum of a number happens to be equal to the remainder obtained when the number is divided by 9. In the case of the answer above—8651—the mod sum was 2. If you divide 8651 by 9, the answer is 961 with a remainder of 2. In other words, if you cast out 9 from 8651 a total of 961 times, you'll have a remainder of 2. There's one small exception to this. Recall that the sum of the digits of any multiple of 9 is also a multiple of 9. Thus, if a number is a multiple of 9, it will have a mod sum of 9, even though it has a remainder of 0.

SUBTRACTING ON PAPER

You can't, of course, subtract columns of numbers the same way you add them. Rather, you subtract them number by number, which means that all subtraction problems involve just two numbers. Once again, with pencil and paper at our disposal, it is easier to subtract from right to left. To check your answer, just add the answer to the second number. If you are correct, then you should get the top number.

If you want, you can also use mod sums to check your

answer. The key is to subtract the mod sums you arrive at and then compare that number to the mod sum of your answer:

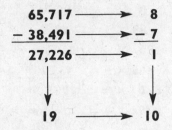

There's one extra twist. If the difference in the mod sums is a negative number or 0, add 9 to it. For instance:

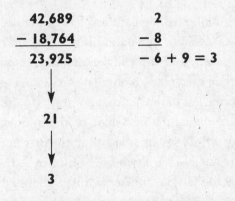

PENCIL-AND-PAPER SQUARE ROOTS

With the advent of pocket calculators, the pencil-and-paper method of calculating square roots has practically become a lost art. You've already seen how to estimate square roots mentally. Now I'll show you how to do it exactly, using pencil and paper.

Remember how in guesstimating square roots you calculated the square root of nineteen? Let's look at that problem again, this time using a method that will give you the exact square root.

$$
\begin{array}{r}
4.\ 3\ \ 5\ \ 8 \\
\sqrt{19.000000} \\
\end{array}
$$

$$
\begin{array}{rcl}
4^2 & = & 16 \\
8\_ \times \_ & \le & 3\ 00 \\
8\underline{3} \times \underline{3} & = & 2\ 49 \\
86\_ \times \_ & \le & 5100 \\
86\underline{5} \times \underline{5} & = & 4325 \\
870\_ \times \_ & \le & 77500 \\
870\underline{8} \times \underline{8} & = & 69664 \\
\end{array}
$$

I will describe the general method that fits all cases, and illustrate it with the above example.

Step 1. If the number of digits to the left of the decimal point is one, three, five, seven, or any odd number of digits, the first digit of the answer (or quotient) will be the largest number whose square is less than the *first* digit of the original number. If the number of digits to the left of the decimal point is two, four, six, or any even number of digits, the first digit of the quotient will be the largest number whose square is less than the first *two* digits of the dividend. In this case, 19 is a two-digit number, so the first digit of the quotient is the largest number whose square is less than 19. That number is 4. Write the answer above either the first digit of the dividend (if odd) or the second digit of the dividend (if even).

Step 2. Subtract the square of the number in step 1, then bring down two more digits. Since $4^2 = 16$, we subtract $19 - 16 = 3$. We bring down two 0s, leaving 300 as the current remainder.

Step 3. Double the current quotient (ignoring any decimal point), and put a blank space in following it. Here $4 \times 2 = 8$. Put down $8\_ \times \_$ to the left of the current remainder, in this case 300.

Step 4. The next digit of the quotient will be the largest number that can be put in both blanks so that the resulting multiplication problem is less than or equal to the current remainder. In this case the number is 3, because 8<u>3</u> × <u>3</u> = 249, whereas 84 × 4 = 336, which is too high. Write this number above the second digit of the next two numbers; in this case the 3 would go above the second 0. We now have a quotient of 4.3.

Step 5. If you want more digits, subtract the product from the remainder (i.e., 300 − 249 = 51), and bring down the next two digits; in this case 51 turns into 5100, which becomes the current remainder. Now repeat steps 3 and 4.

To get the third digit of the square root, double the quotient, again ignoring the decimal point (i.e., 43 × 2 = 86). Place 86_ × _ to the left of 5100. The number 5 gives us 86<u>5</u> × <u>5</u> = 4325, the largest product below 5100. The 5 goes above the next two numbers, in this case two more 0s. We now have a quotient of 4.35. For even more digits, repeat the process as we've done in the example.

Here's an example of an odd number of digits before the decimal point:

$$
\begin{array}{r}
2\,8.\;9\;7 \\
\sqrt{839.4000} \\
2^2 = \underline{4} \\
4\_\;\times\;\_\;\le 439 \\
4\underline{8} \times \underline{8} = \underline{384} \\
56\_\;\times\;\_\;\le 55\,40 \\
56\underline{9} \times \underline{9} = \underline{51\,21} \\
578\_\;\times\;\_\;\le\;4\,1900 \\
578\underline{7} \times \underline{7} = \;4\,0509
\end{array}
$$

Next, we'll calculate the square root of a four-digit number. In this case—as with two-digit numbers—we consider the first two digits of the problem to determine the first digit of the square root:

$$
\begin{array}{r}
\quad\quad\quad 8\ 2.\ 0\ 6 \\
\sqrt{6735.0000} \\
8^2 = \underline{64}\quad\quad\quad\quad \\
16\_ \times \_ \le \ 335\quad\quad \\
16\underline{2} \times \underline{2} = \underline{324}\quad\quad \\
164\_ \times \_ \le \ \ 11\ 00\quad \\
1640 \times \underline{0} = \underline{\quad\quad 0}\quad \\
1640\_ \times \_ \le \ \ 11\ 0000 \\
1640\underline{6} \times \underline{6} = \quad 9\ 8436
\end{array}
$$

Finally, if the number for which you are calculating the square root is a perfect square, you will know it as soon as you end up with a remainder of 0 and nothing to bring down. For example:

$$
\begin{array}{r}
\quad\quad\quad 3.\ 3 \\
\sqrt{10.89} \\
3^2 = \underline{9}\quad\quad \\
6\_ \times \_ \le 1\ 89 \\
6\underline{3} \times \underline{3} = \underline{1\ 89} \\
0
\end{array}
$$

PENCIL-AND-PAPER MULTIPLICATION

For pencil-and-paper multiplication I use the *criss-cross method*, which enables me to write down the entire answer on one line without ever writing any partial results! This is one of the most impressive displays of mathemagics when you have pencil and paper at your disposal. Many lightning calculators in the past earned their reputations with this method. They would be given two large numbers and write down the answer almost instantaneously. The criss-cross method is best learned by example.

$$
\begin{array}{r}
47 \\
\times\ \ 34 \\
\hline
1598
\end{array}
$$

Step 1. First, multiply 4 × 7 to yield 2<u>8</u>, write down the 8, and mentally carry the 2 to the next computation, below.

Step 2. Following the diagram, add 2 + (4 × 4) + (3 × 7) = 3<u>9</u>, write down the 9, and carry the 3 to the final computation, below.

Step 3. End by adding 3 + (3 × 4) = <u>15</u> and writing down 15 to arrive at your final answer.

4   7

3   4

You have now written the answer: <u>1598</u>.

Let's solve another 2-by-2 multiplication problem using the criss-cross method:

$$
\begin{array}{r}
83 \\
\times\ 65 \\
\hline
5395
\end{array}
$$

The steps and diagrams appear as follows:

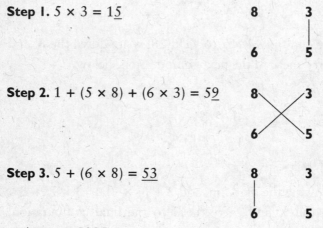

Step 1. $5 \times 3 = 1\underline{5}$

Step 2. $1 + (5 \times 8) + (6 \times 3) = 5\underline{9}$

Step 3. $5 + (6 \times 8) = \underline{53}$

Answer: <u>5395</u>

The criss-cross method gets slightly more complicated with 3-by-3 problems.

$$
\begin{array}{r}
853 \\
\times\ \ \ 762 \\
\hline
649{,}986
\end{array}
$$

We proceed as suggested by our pattern below:

Step 1. $2 \times 3 = \underline{6}$

Step 2. $(2 \times 5) + (6 \times 3) = 2\underline{8}$ 8   5   3

                                             7   6   2

Step 3. $2 + (2 \times 8) + (7 \times 3) + (6 \times 5) = 6\underline{9}$ 8   5   3

                                                              7   6   2

Step 4. $6 + (6 \times 8) + (7 \times 5) = 8\underline{9}$ 8   5   3

                                             7   6   2

Step 5. $8 + (8 \times 7) = \underline{64}$ 8   5   3

                                     7   6   2

Answer: $\underline{649,986}$

Notice that the number of multiplications at each step is 1, 2, 3, 2, and 1 respectively. The mathematics underlying the criss-cross method is nothing more that the distributive law. For instance, $853 \times 762 = (800 + 50 + 3) \times (700 + 60 + 2) = (3 \times 2) + [(5 \times 2) + (3 \times 6)] \times 10 + [(8 \times 2) + (5 \times 6) + (3 \times 7)] \times 100 + [(8 \times 6) + (5 \times 7)] \times 1000 + (8 \times 7) \times 10,000$, which are precisely the calculations of the criss-cross method.

You can check your answer with the mod sum method by multiplying the mod sums of the two numbers and computing the resulting number's mod sum. Compare this number to the mod sum of the answer. If your answer is correct, they must match. For example:

$$
\begin{array}{cc}
853 & 7 \\
\times \ 762 & \times\ 6 \\
\hline
649{,}986 & 42 \\
\downarrow & \downarrow \\
42 & 6 \\
\downarrow & \\
6 &
\end{array}
$$

If the mod sums don't match, you made a mistake. This method will detect mistakes, on average, about 8 times out of 9.

In the case of 3-by-2 multiplication problems, the procedure is the same except you treat the hundreds digit of the second number as a 0:

$$
\begin{array}{c}
846 \\
\times\ 037 \\
\hline
31{,}302
\end{array}
$$

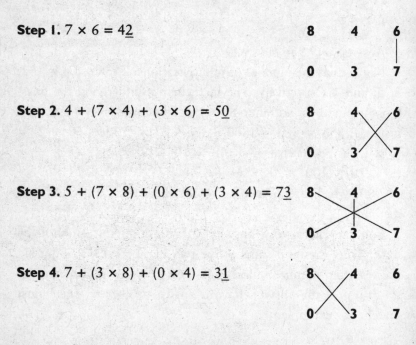

Step 1. $7 \times 6 = 4\underline{2}$

    8   4   6

    0   3   7

Step 2. $4 + (7 \times 4) + (3 \times 6) = 5\underline{0}$

Step 3. $5 + (7 \times 8) + (0 \times 6) + (3 \times 4) = 7\underline{3}$

Step 4. $7 + (3 \times 8) + (0 \times 4) = 3\underline{1}$

Step 5. $3 + (0 \times 8) = \underline{3}$

$$\begin{array}{ccc} 8 & 4 & 6 \\ \vert & & \\ 0 & 3 & 7 \end{array}$$

Answer: <u>31,302</u>

Of course, you would normally just ignore the multiplication by zero, in practice.

You can use the criss-cross to do any size multiplication problem. To answer the 5-by-5 problem below will require nine steps. The number of multiplications in each step is 1, 2, 3, 4, 5, 4, 3, 2, 1, for twenty-five multiplications altogether.

$$\begin{array}{r} 42,867 \\ \times \quad 52,049 \\ \hline 2,231,184,483 \end{array}$$

Step 1. $9 \times 7 = 6\underline{3}$

$$\begin{array}{ccccc} 4 & 2 & 8 & 6 & 7 \\ & & & & \vert \\ 5 & 2 & 0 & 4 & 9 \end{array}$$

Step 2. $6 + (9 \times 6) + (4 \times 7) = 8\underline{8}$

$$\begin{array}{ccccc} 4 & 2 & 8 & 6 & 7 \\ & & & & \\ 5 & 2 & 0 & 4 & 9 \end{array}$$

Step 3. $8 + (9 \times 8) + (0 \times 7) + (4 \times 6) = 10\underline{4}$

$$\begin{array}{ccccc} 4 & 2 & 8 & 6 & 7 \\ & & & & \\ 5 & 2 & 0 & 4 & 9 \end{array}$$

Step 4. $10 + (9 \times 2) + (2 \times 7) + (4 \times 8) + (0 \times 6) = 7\underline{4}$

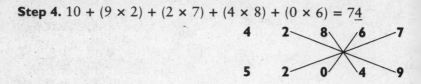

Step 5. $7 + (9 \times 4) + (5 \times 7) + (4 \times 2) + (2 \times 6) + (0 \times 8) = 9\underline{8}$

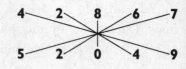

Step 6. $9 + (4 \times 4) + (5 \times 6) + (0 \times 2) + (2 \times 8) = 7\underline{1}$

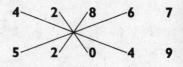

Step 7. $7 + (0 \times 4) + (5 \times 8) + (2 \times 2) = 5\underline{1}$

Step 8. $5 + (2 \times 4) + (5 \times 2) = 2\underline{3}$

Step 9. $(5 \times 4) + 2 = \underline{\underline{22}}$

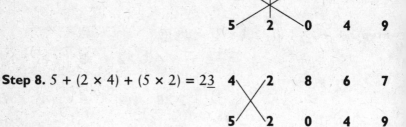

Answer: <u>2,231,184,483</u>
You can check your answer by using the mod sums method.

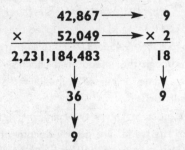

CASTING OUT ELEVENS

To double-check your answer another way, you can use the method known as casting out elevens. It's similar to casting out nines, except you reduce the numbers by alternately subtracting and adding the digits from right to left, ignoring any decimal point. If the result is negative, then add eleven to it. (It may be tempting to do the addition and subtraction from left to right, as you do with mod sums, but in this case you must do it from right to left for it to work.)
For example:

$$234.87 \longrightarrow 7 - 8 + 4 - 3 + 2 = 2 \longrightarrow 2$$
$$+\ 58.61 \longrightarrow 1 - 6 + 8 - 5 = \underline{-2} \longrightarrow \underline{9}$$
$$293.48 \longrightarrow 8 - 4 + 3 - 9 + 2 = 0 \longrightarrow 11 \longrightarrow 0$$

The same method works for subtraction problems:

$$65,717 \longrightarrow 14 \longrightarrow 3$$
$$-\ 38,491 \longrightarrow -(-\ 9) \longrightarrow \underline{+9}$$
$$27,226 \longrightarrow 12 \longrightarrow 1$$
$$1$$

Shakuntala Devi: That's Incalculable!

In 1976 the *New York Times* reported that an Indian woman named Shakuntala Devi (b. 1939) added 25,842 + 111,201,721 + 370,247,830 + 55,511,315, and then multiplied that sum by 9,878, for a correct total of 5,559,369,456,432, all in less than twenty seconds. Hard to believe, though this uneducated daughter of impoverished parents has made a name for herself in the United States and Europe as a lightning calculator.

Unfortunately, most of Devi's truly amazing feats not done by obvious "tricks of the trade" are poorly documented. Her greatest claimed accomplishment—the timed multiplication of two thirteen-digit numbers on paper—has appeared in the *Guinness Book of World Records* as an example of a "Human Computer." The time of the calculation, however, is questionable at best. Devi, a master of the criss-cross method, reportedly multiplied 7,686,369,774,870 × 2,465,099,745,799, numbers randomly generated at the computer department of Imperial College, London, on June 18, 1980. Her correct answer of 18,947,668,177,995,426,773,730 was allegedly generated in an incredible twenty seconds. *Guinness* offers this disclaimer: "Some eminent mathematical writers have questioned the conditions under which this was apparently achieved and predict that it would be impossible for her to replicate such a feat under highly rigorous surveillance." Since she had to calculate 169 multiplication problems and 167 addition problems, for a total of 336 operations, she would have had to do each calculation in under a tenth of a second, with no mistakes, taking the time to write down all 26 digits of the answer. Reaction time alone makes this record truly in the category of "that's incalculable!"

Despite this, Devi has proven her abilities doing rapid calculation and has even written her own book on the subject.

It even works with multiplication:

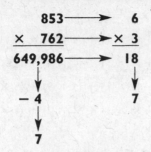

$$
\begin{array}{rcl}
853 & \longrightarrow & 6 \\
\times\ 762 & \longrightarrow & \times\ 3 \\
\hline
649,986 & \longrightarrow & 18 \\
\downarrow & & \downarrow \\
-4 & & 7 \\
\downarrow & & \\
7 & &
\end{array}
$$

If the numbers disagree, you made a mistake somewhere. But if they match, it's still possible for a mistake to exist. On average, this method will detect an error 10 times out of 11. Thus a mistake has a 1 in 11 chance of sneaking past the elevens check, a 1 in 9 chance of sneaking past the nines check, and only a 1 in 99 chance of being undetected if both checks are used. For more information about this and other fascinating mathemagical topics, I would encourage you to read any of Martin Gardner's recreational math books.

You are now ready for the ultimate pencil-and-paper multiplication problem in the book, a 10-by-10! This has no practical value whatsoever except for showing off! (And personally I think multiplying five-digit numbers is impressive enough since the answer will be at the capacity of most calculators.) We present it here just to prove that it can be done. The criss-crosses follow the same basic pattern as that in the 5-by-5 problem. There will be nineteen computation steps and at the tenth step there will be ten criss-crosses! Here you go:

$$
\begin{array}{r}
2,766,829,451 \\
\times\ 4,425,575,216
\end{array}
$$

Here's how we do it:

Step 1. $6 \times 1 = \underline{6}$

Step 2. $(6 \times 5) + (1 \times 1) = 3\underline{1}$

Step 3. $3 + (6 \times 4) + (2 \times 1) + (1 \times 5) = 3\underline{4}$

Step 4. $3 + (6 \times 9) + (5 \times 1) + (1 \times 4) + (2 \times 5) = 7\underline{6}$

Step 5. $7 + (6 \times 2) + (7 \times 1) + (1 \times 9) + (5 \times 5) + (2 \times 4) = 6\underline{8}$

Step 6. $6 + (6 \times 8) + (5 \times 1) + (1 \times 2) + (7 \times 5) + (2 \times 9) + (5 \times 4) = 13\underline{4}$

Step 7. $13 + (6 \times 6) + (5 \times 1) + (1 \times 8) + (5 \times 5) + (2 \times 2) + (7 \times 4) + (5 \times 9) = 16\underline{4}$

Step 8. $16 + (6 \times 6) + (2 \times 1) + (1 \times 6) + (5 \times 5) + (2 \times 8) + (5 \times 4) + (5 \times 2) + (7 \times 9) = 19\underline{4}$

Step 9. $19 + (6 \times 7) + (4 \times 1) + (1 \times 6) + (2 \times 5) + (2 \times 6) + (5 \times 4) + (5 \times 8) + (5 \times 9) + (7 \times 2) = 21\underline{2}$

Step 10. $21 + (6 \times 2) + (4 \times 1) + (1 \times 7) + (4 \times 5) + (2 \times 6) + (2 \times 4) + (5 \times 6) + (5 \times 9) + (7 \times 8) + (5 \times 2) = 22\underline{5}$

Step 11. $22 + (1 \times 2) + (4 \times 5) + (2 \times 7) + (4 \times 4) + (5 \times 6) + (2 \times 9) + (7 \times 6) + (5 \times 2) + (5 \times 8) = 21\underline{4}$

Step 12. $21 + (2 \times 2) + (4 \times 4) + (5 \times 7) + (4 \times 9) + (7 \times 6) + (2 \times 2) + (5 \times 6) + (5 \times 8) = 22\underline{8}$

Step 13. $22 + (5 \times 2) + (4 \times 9) + (7 \times 7) + (4 \times 2) + (5 \times 6) + (2 \times 8) + (5 \times 6) = 20\underline{1}$

Step 14. $20 + (7 \times 2) + (4 \times 2) + (5 \times 7) + (4 \times 8) + (5 \times 6) + (2 \times 6) = 15\underline{1}$

Step 15. $15 + (5 \times 2) + (4 \times 8) + (5 \times 7) + (4 \times 6) + (2 \times 6) = 12\underline{8}$

Step 16. $12 + (5 \times 2) + (4 \times 6) + (2 \times 7) + (4 \times 6) = 8\underline{4}$

Step 17. $8 + (2 \times 2) + (4 \times 6) + (4 \times 7) = 6\underline{4}$

Step 18. $6 + (4 \times 2) + (4 \times 7) = 4\underline{2}$

Step 19. $4 + (4 \times 2) = \underline{12}$

If you were able to negotiate this extremely difficult problem successfully the first time through, you are on the verge of graduating from apprentice to master mathemagician!

$$
\begin{array}{r}
2,766,829,451 \longrightarrow 5 \\
\times\ 4,425,575,216 \longrightarrow 5 \\
\hline
12,244,811,845,244,486,416 \longrightarrow 7
\end{array}
$$

PENCIL-AND-PAPER MATHEMATICS EXERCISES

EXERCISE: COLUMNS OF NUMBERS

Sum the following column of numbers, checking your answer by reading the column from bottom to top, and then by the mod sums method. If the two mod sums do not match, recheck your addition:

1.
$$
\begin{array}{r}
672 \\
1367 \\
107 \\
7845 \\
358 \\
210 \\
+\ 916 \\
\end{array}
$$

Sum this column of dollars and cents:

2.
$$
\begin{array}{r}
\$\ 21.56 \\
19.38 \\
211.02 \\
9.16 \\
26.17 \\
+\ \ 1.43 \\
\end{array}
$$

EXERCISE: SUBTRACTING ON PAPER

Subtract the following numbers, using mod sums to check your answer and then by adding the bottom two numbers to get the top number:

1. 75,423
    − 46,298

2. 876,452
    − 593,876

3. 3,249,202
    − 2,903,445

4. 45,394,358
    − 36,472,659

EXERCISE: SQUARE ROOT GUESSTIMATION

For the following numbers, compute the exact square root using the doubling and dividing technique.

1. $\sqrt{15}$ 2. $\sqrt{502}$ 3. $\sqrt{439.2}$ 4. $\sqrt{361}$

EXERCISE: PENCIL-AND-PAPER MULTIPLICATION

To wrap up this exercise session, use the criss-cross method to compute exact multiplication problems of any size. Place the answer below the problems on one line, from right to left. Check your answers using mod sums.

1. 54
    × 37

2. 273
    × 217

3. 725
    × 609

4. 3,309
    × 2,868

5. 52,819
    × 47,820

6. 3,923,759
    × 2,674,093

Chapter 7

A Memorable Chapter: Memorizing Numbers

One of the most common questions I am asked pertains to my memory. No, I don't have an extraordinary memory. Rather, I apply a memory system that can be learned by anyone and is described in the following pages. In fact, experiments have shown that almost anyone of average intelligence can be taught to vastly improve their ability to memorize numbers.

In this chapter we'll teach you a way to improve your memory for numbers. Not only does this have obvious practical benefits, such as remembering dates or recalling phone numbers, but it also allows the mathemagician to solve much larger problems mentally. In the next chapter we'll show you how to apply the techniques of this chapter to multiply five-digit numbers in your head!

USING MNEMONICS

The method presented here is an example of a mnemonic—that is, a tool to improve memory encoding and retrieval. Mnemonics work by converting incomprehensible data (such as digit sequences) to something more meaningful. For example, take just a moment to memorize the sentence below:

"My turtle Pancho will, my love, pick up my new mover, Ginger."

Read it over several times. Look away from the page and say it to yourself over and over until you don't have to look back at the page, visualizing the turtle, Pancho, picking up your new mover, Ginger, as you do so. Got it?

Congratulations! You have just memorized the first twenty-four digits of the mathematical expression π (pi). Recall that π is the ratio of the circumference of a circle to its diameter, usually approximated in school as 3.14, or $\frac{22}{7}$. In fact, π is an irrational number (one whose digits continue indefinitely with no repetition or pattern), and computers have been used to calculate π to billions of places.

THE PHONETIC CODE

I'm sure you're wondering how "My turtle Pancho will, my love, pick up my new mover, Ginger" translates into 24 places of π.

To do this, you first need to memorize the phonetic code below, where each number between 0 and 9 is assigned a consonant sound.

1 is the *t* or *d* sound.

2 is the *n* sound.

3 is the *m* sound.

4 is the *r* sound.

5 is the *l* sound.

6 is the *j*, *ch*, or *sh* sound.

7 is the *k* or hard *g* sound.

8 is the *f* or *v* sound.

9 is the *p* or *b* sound.

0 is the *z* or *s* sound.

Memorizing this code isn't as hard as it looks. For one thing, notice that in cases where more than one letter is associated with a number, they have similar pronunciations. For example, the *k* sound (as it appears in words like *kite* or *cat*) is similar to the hard *g* sound (as it appears in such words as *goat*). You can also rely on the following tricks to help you memorize the code:

1 A typewritten *t* or *d* has just 1 downstroke.

2 A typewritten *n* has 2 downstrokes.

3 A typewritten *m* has 3 downstrokes.

4 The number 4 ends in the letter *r*.

5 Shape your hand with 4 fingers up and the thumb at a 90-degree angle—that's 5 fingers in an L shape.

6 A *J* looks sort of like a backward 6.

7 A K can be drawn by laying two 7s back to back.

8 A lowercase *f* written in cursive looks like an 8.

9 The number 9 looks like a backward *p* or an upside-down *b*.

0 The word *zero* begins with the letter *z*.

Or you can just remember the list in order, by thinking of the name Tony Marloshkovips!

Practice remembering this list. In about ten minutes you should have all the one-digit numbers associated with consonant sounds. Next, you can convert numbers into words by placing vowel sounds around or between the consonant sounds. For instance, the number 32 can become any of the following words: *man, men, mine, mane, moon, many, money, menu, amen, omen, amino, mini, minnie,* and so on. Notice that the word *minnie* is acceptable since the *n* sound is only used once.

The following words could *not* represent the number 32 because they use other consonant sounds: *mourn, melon, mint.* These words would be represented by the numbers 342, 352, and 321, respectively. The consonant sounds of *h, w,* and *y* can be added freely since they don't appear on the list. Hence, 32 can also become *human, woman, yeoman,* or *my honey.*

The following list gives you a good idea of the types of words you can create using this phonetic code. Don't feel obligated to memorize it—use it as inspiration to explore the possibilities on your own.

1	tie	19	tub	37	mug	55	lily
2	knee	20	nose	38	movie	56	leash
3	emu	21	nut	39	map	57	log
4	ear	22	nun	40	rose	58	leaf
5	law	23	name	41	rod	59	lip
6	shoe	24	Nero	42	rain	60	cheese
7	cow	25	nail	43	ram	61	sheet
8	ivy	26	notch	44	rear	62	chain
9	bee	27	neck	45	roll	63	chum
10	dice	28	knife	46	roach	64	cherry
11	tot	29	knob	47	rock	65	jail
12	tin	30	mouse	48	roof	66	judge
13	tomb	31	mat	49	rope	67	chalk
14	tire	32	moon	50	lace	68	chef
15	towel	33	mummy	51	light	69	ship
16	dish	34	mower	52	lion	70	kiss
17	duck	35	mule	53	lamb	71	kite
18	dove	36	match	54	lure	72	coin

73	comb	80	face	87	fog	94	bear
74	car	81	fight	88	fife	95	bell
75	coal	82	phone	89	V.I.P.	96	beach
76	cage	83	foam	90	bus	97	book
77	cake	84	fire	91	bat	98	puff
78	cave	85	file	92	bun	99	puppy
79	cap	86	fish	93	bomb	100	daisies

The Number-Word List

For practice, translate the following numbers into words, then check the correct translation below. When translating numbers into words, there are a variety of words that can be formed:

42

74

67

86

93

10

55

826

951

620

367

Here are some words that I came up with:

42: rain, rhino, Reno, ruin, urn

74: car, cry, guru, carry

67: jug, shock, chalk, joke, shake, hijack

86: fish, fudge

93:   bum, bomb, beam, palm, pam

10:   toss, dice, toes, dizzy, oats, hats

55:   lily, lola, hallelujah!

826:   finch, finish, vanish

951:   pilot, plot, belt, bolt, bullet

620:   jeans, chains, genius

367:   magic!

As an exercise, translate each of the following words into its unique number:

dog

oven

cart

fossil

banana

garage

pencil

Mudd

multiplication

Cleveland

Ohio

Answers:

dog:	17
oven:	82
cart:	741
fossil:	805
banana:	922
garage:	746
pencil:	9,205

Mudd:	31
multiplication:	35,195,762
Cleveland:	758,521
Ohio:	(nothing)

Although a number can usually be converted into many words, a word can be translated only into a single number. This is an important property for our applications as it enables us to recall specific numbers.

Using this system you can translate any number or series of numbers (e.g., phone numbers, Social Security numbers, driver's license numbers, the digits of π) into a word or a sentence. Here's how the code works to translate the first twenty-four digits of π:

3 1415 926 5 3 58 97 9 3 2 384 6264
"My turtle Pancho will, my love, pick up my new mover, Ginger."

Remember that, with this phonetic code, *g* is a hard sound, as in *grass,* so a soft *g* (as in *Ginger*) sounds like *j* and is represented by a 6. Also, the word *will* is, phonetically, just *L,* and is represented by 5, since the consonant sound *w* can be used freely. Since this sentence can only be translated back to the twenty-four digits above, you have effectively memorized π to twenty-four digits!

There's no limit to the number of numbers this code will allow you to memorize. For example, the following two sentences, when added to "My turtle Pancho will, my love, pick up my new mover, Ginger," will allow you to memorize the first sixty digits of π:

3 38 327 950 2 8841 971
"My movie monkey plays in a favorite bucket."

And:

> 69 3 99   375 1 05820   97494
> **"Ship my puppy Michael to Sullivan's backrubber."**

What the heck, let's go for a hundred digits:

> 45   92   307 81   640   62 8 620
> **"A really open music video cheers Jenny F. Jones."**

> 8   99 86 28 0 3482 5   3421   1   7067
> **"Have a baby fish knife so Marvin will marinate the goosechick."**

You can really feel proud of yourself once these sentences roll trippingly off your tongue, and you're able to translate them quickly back into numbers. But you've got a ways to go for the world record. Hiroyuki Goto of Japan recited π to 42,195 places, from memory, in seventeen hours and twenty-one minutes in 1995.

HOW MNEMONICS MAKES MENTAL CALCULATION EASIER

Aside from improving your ability to memorize long sequences of numbers, mnemonics can be used to store partial results in the middle of a difficult mental calculation. For example, here's how you can use mnemonics to help you square a three-digit number:

A Piece of Pi for Alexander Craig Aitken

Perhaps one of the most impressive feats of mental calculation was performed by a professor of mathematics at the University of Edinburgh, Alexander Craig Aitken (1895–1967), who not only learned the value of π to 1,000 places but, when asked to demonstrate his amazing memory during a lecture, promptly rattled off the first 250 digits of π. He was then challenged to skip ahead and begin at the 551st digit and continue for another 150 places. This he did successfully without a single error.

How did he do it? Aitken explained to his audience that "the secret, to my mind, is relaxation, the complete antithesis of concentration as usually understood." Aitken's technique was more auditory than usual. He arranged the numbers into chunks of fifty digits and memorized them in a sort of rhythm. With undaunted confidence he explained, "It would have been a reprehensibly useless feat had it not been so easy."

That Aitken could memorize π to a thousand places does not qualify him as a lightning calculator. That he could easily multiply five-digit numbers against each other does. A mathematician named Thomas O'Beirne recalled Aitken at a desk calculator demonstration. "The salesman," O'Beirne wrote, "said something like 'We'll now multiply 23,586 by 71,283.' Aitken said right off, 'And get . . .' (whatever it was). The salesman was too intent on selling even to notice, but his manager, who was watching, did. When he saw Aitken was right, he nearly threw a fit (and so did I)."

Ironically, Aitken noted that when he bought a desk calculator for himself, his own mental skills deteriorated quickly. Anticipating what the future might hold, he lamented, "Mental calculators may, like the Tasmanian or the Maori, be doomed to extinction. Therefore you may be able to feel an almost anthropological interest in surveying a curious specimen, and some of my auditors here may be able to say in the year A.D. 2000, 'Yes, I knew one such.'" Fortunately, history has proved him wrong!

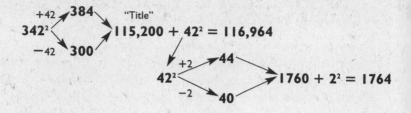

As you recall from Chapter 3, to square 342 you first multiply 300 × 384, for 115,200, then add 42^2. But by the time you've computed $42^2 = 1,764$, you may have forgotten the earlier number, 115,200. Here's where our memory system comes to the rescue. To store the number 115,200, put 200 on your hand by raising two fingers, and convert 115 into a word like *title*. (By the way, I do not consider storing the 200 on fingers to be cheating. After all, what are hands for if not for holding on to digits!) Repeat the word *title* to yourself once or twice. That's easier to remember than 115,200, especially after you start calculating 42^2. Once you've arrived at 1,764, you can add that to *title 2,* or 115,200, for a total of 116,964.

Here's another:

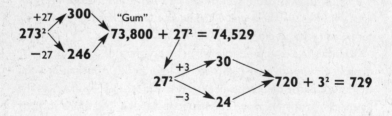

After multiplying 300 × 246 = 73,800, convert 73 into *gum* and hold 800 on your hand by raising eight fingers. Once you've computed $27^2 = 729$, just add that to *gum 8,* or 73,800, for a total of 74,529. This may seem a bit cumbersome at first, but with practice the conversion from numbers to words and back to numbers becomes almost second nature.

You have seen how easily two-digit numbers can be translated into simple words. Not all three-digit numbers can be translated so easily, but if you're at a loss for a simple word to act as a mnemonic, an unusual word or a nonsense word will do. For example, if no simple word for 286 or 638 comes quickly to mind, use a combination word like *no fudge* or a nonsense word like *jam-off*. Even these unusual words should be easier to recall than 286 or 638 during a long calculation. For some of the huge problems in the next chapter, these memory tricks will be indispensable.

MEMORY MAGIC

Without using mnemonics, the average human memory (including mine) can hold only about seven or eight digits at a time. But once you've mastered the ability to change numbers into words, you can expand your memory capacity considerably. Have someone slowly call out sixteen digits while someone else writes them down on a blackboard or a piece of paper. Once they are written down, you repeat them back in the exact order they were given without looking at the board or the piece of paper! At a recent lecture demonstration, I was given the following series of numbers:

1, 2, 9, 7, 3, 6, 2, 7, 9, 3, 3, 2, 8, 2, 6, 1

As the numbers were called out, I used the phonetic code to turn them into words and then tied them together with a nonsensical story. In this case, 12 becomes *tiny*, 97 becomes *book*,

362 becomes *machine*, 793 becomes *kaboom*, 32 becomes *moon*, and 8261 becomes *finished*.

As the words were being created, I linked them together to form a silly story to help me remember them. I pictured finding a *tiny book*, which I placed inside a *machine*. This caused the machine to go *kaboom*, and tossed me to the *moon*, where I was *finished*. This story may sound bizarre, but the more ridiculous the story, the easier it is to remember—and besides, it's more fun.

Chapter 8

The Tough Stuff Made Easy: Advanced Multiplication

At this point in the book—if you've gone through it chapter by chapter—you have learned to do mental addition, subtraction, multiplication, and division, as well as the art of guesstimation, pencil-and-paper mathemagics, and the phonetic code for number memory. This chapter is for serious, die-hard mathemagicians who want to stretch their minds to the limits of mental calculation. The problems in this chapter range from four-digit squares to the largest problem I perform publicly—the multiplication of two different five-digit numbers.

In order to do these problems, it is particularly important that you are comfortable and reasonably fast using the phonetic code. And even though, if you glance ahead in this chapter, the problems seem overwhelming, let me restate the two fundamental premises of this book: (1) all mental calculation skills can be learned by almost anyone; and (2) the key is the simplification of all problems into easier problems that can be done quickly. There is no problem in this chapter (or anywhere else) that you will encounter (comparable to these) that cannot be mastered and learned through the simplification techniques you've learned in previous chapters. Because we are assuming that you've mastered the techniques needed for this chapter, we will be teaching primarily by illustration, rather than walking you through the problems word for word. As an aid, however, we remind you that many of the simpler problems embedded within these larger problems are ones you have already encountered in previous chapters.

We begin with four-digit squares. Good luck!

FOUR-DIGIT SQUARES

As a preliminary skill for mastering four-digit squares, you need to be able to do 4-by-1 problems. We break these down into two 2-by-1 problems, as in the examples below:

$$
\begin{array}{r}
\textbf{4,867 (4,800 + 67)} \\
\times \qquad \textbf{9} \\
\hline
9 \times 4,800 = \quad \textbf{43,200} \\
9 \times 67 = +\quad \textbf{603} \\
\hline
\textbf{43,803}
\end{array}
$$

$$
\begin{array}{r}
\textbf{2,781 (2,700 + 81)} \\
\times \qquad \textbf{4} \\
\hline
4 \times 2,700 = \quad \textbf{10,800} \\
4 \times 81 = +\quad \textbf{324} \\
\hline
\textbf{11,124}
\end{array}
$$

$$
\begin{array}{r}
\textbf{6,718 (6,700 + 18)} \\
\times \qquad \textbf{8} \\
\hline
8 \times 6,700 = \quad \textbf{53,600} \\
8 \times 18 = +\quad \textbf{144} \\
\hline
\textbf{53,744}
\end{array}
$$

$$4,269 \ (4,200 + 69)$$
$$\times \ \ \ \ \ 5$$

$$5 \times 4,200 = \ \ 21,000$$
$$5 \times 69 = \ + \ \ \ \ 345$$
$$21,345$$

Once you've mastered 4-by-1s you're ready to tackle four-digit squares. Let's square 4,267. Using the same method we used with two-digit and three-digit squares, we'll square the number 4,267 by rounding down by 267 to 4,000, and up by 267 to 4,534, multiplying 4,534 × 4,000 (a 4-by-1), and then adding the square of the number you went up and down by, or 267^2, as illustrated below:

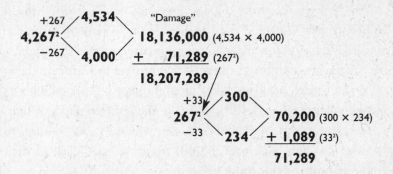

Now, obviously there is a lot going on in this problem. I realize it's one thing to say, "Just add the square of 267" and quite another to actually do it and remember what it was you were supposed to add to it. First of all, once you multiply 4,534 × 4 to get 18,136, you can actually say the first part of the answer out loud: "Eighteen million . . ." You can do so because the original number is always rounded to the nearest thousand. Thus the largest three-digit number you will ever square in the next step is 500. The square of 500 is 250,000, so as soon as you see that the rest of your answer (in this case, 136,000) is less than 750,000, you know that the millions digit will not change.

Once you've said "eighteen million . . . ," you need to hold on to 136,000 before you square 267. Here's where our mnemonics from the last chapter comes to the rescue! Using the phonetic code, you can convert 136 to *damage* (1 = *d*, 3 = *m*, 6 = *j*). Now you can work on the next part of the problem and just remember *damage* (and that there are three zeros following—this will always be the case). If at any time in the computations you forget what the original problem is, you can either glance at the original number or, if it isn't written down, ask the audience to repeat the problem (which gives the illusion you are starting the problem over from scratch when, in fact, you have already done some of the calculations)!

You now do the three-digit square just as you learned to do before, to get 71,289. I used to have trouble remembering the hundreds digits of my answer (2, in this case). I cured this by raising my fingers (two fingers here). If you forget the last two digits (89), you can go back to the original number (4,267), square its last two digits ($67^2 = 4,489$), and take the last two digits of that.

To compute the final answer, you now add 71,289 to *damage* (which translates back to 136,000) resulting in 207,289, which you may now say out loud.

Let's do one more four-digit square—$8,431^2$:

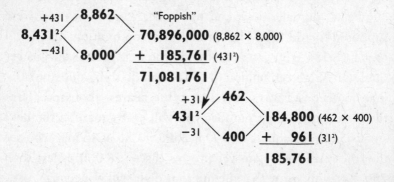

Without repeating all the detailed steps as we did in the last problem, I will point out some of the highlights of this problem. After doing the $8 \times 8{,}862 = 70{,}896$, note that 896 is above 750, so a carry is possible. In fact, since 431^2 is greater than $400^2 = 160{,}000$, there definitely will be a carry when you do the addition to the number 896,000. Hence we can safely say aloud, "Seventy-one million . . ." at this point.

When you square 431, you get 185,761. Add the 185 to the 896 to get 1,081, and say the rest of the answer. But remember that you already anticipated the carry, so you just say, ". . . 81 thousand . . . 761." You're done!

We illustrate one more fine point with $2{,}753^2$:

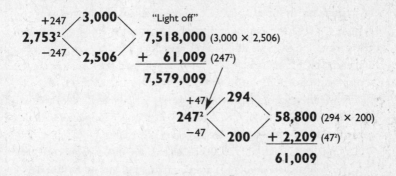

Since you are rounding up to 3,000, you will be multiplying 3,000 by another number in the 2,000s. You could subtract $2{,}753 - 247 = 2{,}506$, but that's a little messy. To obtain the last three digits, double 753 to get 1,506. The last three digits of this number, 506, are the last three digits of the 2,000 number: 2,506! This works because the two numbers being multiplied must add to twice the original number.

Then proceed in the usual manner of multiplying $3{,}000 \times 2{,}506 = 7{,}518{,}000$, convert 518 to the words *light off*, and say

Thomas Fuller: Learned Men and Great Fools

It would be hard to top the physical handicap on learning experienced by Helen Keller, though the social handicap imposed on Thomas Fuller, born in Africa in 1710, comes close. He was not only illiterate; he was forced to work in the fields of Virginia as a slave and never received a single day of education. The "property" of Mrs. Elizabeth Cox, Thomas Fuller taught himself to count to 100, after which he increased his numerical powers by counting such items at hand as the grains in a bushel of wheat, the seeds in a bushel of flax, and the number of hairs in a cow's tail (2,872).

Extrapolating from mere counting, Fuller learned to compute the number of shingles a house would need to cover its roof, the number of posts needed to enclose it, and other numbers relevant to materials needed in construction. His prodigious skills grew, and with them his reputation. In his old age, two Pennsylvanians challenged him to compute, in his head, numbers that would challenge the best lightning calculators. For example, they asked: "Suppose a farmer has six sows, and each sow has six female pigs the first year, and they all increase in the same proportion, to the end of eight years, how many sows will the farmer then have?" The problem can be written as $7^8 \times 6$, that is, $(7 \times 7 \times 7 \times 7 \times 7 \times 7 \times 7 \times 7) \times 6$. Within ten minutes Fuller gave his response of 34,588,806, the correct answer.

Upon Fuller's death in 1790, the *Columbian Centinel* reported that "he could give the number of poles, yards, feet, inches, and barleycorns in any given distance, say the diameter of the earth's orbit; and in every calculation he would produce the true answer, in less time than ninety-nine men in a hundred would take with their pens." When Fuller was asked if he regretted having never gained a traditional education he responded: "No, Massa—it is best I got no learning: for many learned men be great fools."

the first part of the answer out loud as "Seven million . . ." You can say this with confidence since 518 is below 750, so there will not be a carry.

Next, you add the square of 247. Don't forget that you can

derive 247 quickly as the complement of 753. Then proceed to the final answer as you did in the previous four-digit problems.

EXERCISE: FOUR-DIGIT SQUARES

1. 1234^2 　　2. 8639^2 　　3. 5312^2
4. 9863^2 　　5. 3618^2 　　6. 2971^2

3-BY-2 MULTIPLICATION

We saw in the 2-by-2 multiplication problems that there are several different ways to tackle the same problem. The variety of methods increases when you increase the number of digits in the problem. In 3-by-2 multiplication I find that it pays to take a few moments to look at the problem to determine the method of calculation that will put the least amount of strain on the brain.

Factoring Methods

The easiest 3-by-2 problems to compute are those in which the two-digit number is factorable.

For example:

$$637$$
$$\times\ 56\ (8 \times 7)$$

$$637 \times 56 = 637 \times 8 \times 7 = 5,096 \times 7 = 35,672$$

These are great because you don't have to add anything. You just factor 56 into 8×7, then do a 3-by-1 ($637 \times 8 = 5,096$), and finally a 4-by-1 ($5,096 \times 7 = 35,672$). There are no addition steps and you don't have to store any partial results.

More than half of all two-digit numbers are factorable into

numbers 11 and below, so you'll be able to use this method for many problems. Here's an example:

$$853$$
$$\times \ \underline{44} \ (11 \times 4)$$

$$853 \times 11 \times 4 = 9{,}383 \times 4 = 37{,}532$$

To multiply 853×11, you treat 853 as $850 + 3$ and proceed as follows:

$$850 \times 11 = \ \ 9{,}350$$
$$3 \times 11 = \underline{+ \ \ \ \ 33}$$
$$9{,}383$$

Now multiply $9{,}383 \times 4$ by treating 9,383 as $9{,}300 + 83$, as follows:

$$9{,}300 \times 4 = \ \ 37{,}200$$
$$83 \times 4 = \underline{+ \ \ \ 332}$$
$$37{,}532$$

If the two-digit number is not factorable into small numbers, examine the three-digit number to see if it can be factored:

$$144 \ (6 \times 6 \times 4)$$
$$\times \ \underline{76}$$

$$76 \times 144 = 76 \times 6 \times 6 \times 4 = 456 \times 6 \times 4$$
$$= 2{,}736 \times 4 = 10{,}944$$

Notice that the sequence of the multiplication problems is a 2-by-1, a 3-by-1, and finally a 4-by-1. Since these are all problems you can now do with considerable ease, this type of 3-by-2 problem should be no problem at all.

Here's another example where the two-digit number is not factorable but the three-digit number is:

$$462 \ (11 \times 7 \times 6)$$
$$\underline{\times \ 53}$$

$$53 \times 11 \times 7 \times 6 = 583 \times 7 \times 6 = 4{,}081 \times 6 = 24{,}486$$

Here the sequence is a 2-by-2, a 3-by-1, and a 4-by-1, though when the three-digit number is factorable by 11, you can use the elevens method for a very simple 2-by-2 ($53 \times 11 = 583$). In this case it pays to be able to recognize when a number is divisible by 11, as described in Chapter 4.

If the two-digit number is not factorable, and the three-digit number is factorable into only a 2-by-1, the problem can still be handled easily by a 2-by-2, followed by a 4-by-1:

$$423 \ (47 \times 9)$$
$$\underline{\times \ 83}$$

$$83 \times 47 \times 9 = 3{,}901 \times 9 = 35{,}109$$

Here you need to see that 423 is divisible by 9, setting up the problem as $83 \times 47 \times 9$. This 2-by-2 is not so easy, but by treating 83 as $80 + 3$, you get:

$$83 \ (80 + 3)$$
$$\underline{\times \quad 47}$$
$$80 \times 47 = \quad 3{,}760$$
$$3 \times 47 = \underline{+ \ 141}$$
$$3{,}901$$

Then do the 4-by-1 problem of 3,901 × 9 for your final answer of 35,109.

The Addition Method

If the two-digit number and the three-digit number cannot be conveniently factored in the 3-by-2 problem you're doing, you can always resort to the addition method:

$$
\begin{array}{r}
721 \ (720 + 1) \\
\times \quad 37 \\
\end{array}
$$

$$
\begin{array}{rl}
720 \times 37 = & 26,640 \text{ (treating 72 as 9 × 8)} \\
1 \times 37 = & +\quad 37 \\
\hline
& 26,677
\end{array}
$$

The method requires you to add a 2-by-2 (times 10) to a 2-by-1. These problems tend to be more difficult than problems that can be factored since you have to perform a 2-by-1 while holding on to a five-digit number, and then add the results together. In fact, it is probably easier to solve this problem by factoring 721 into 103 × 7, then computing 37 × 103 × 7 = 3,811 × 7 = 26,677.

Here's another example:

$$
\begin{array}{r}
732 \ (730 + 2) \\
\times \quad 57 \\
\end{array}
$$

$$
\begin{array}{rl}
730 \times 57 = & 41,610 \text{ (treating 73 as 70 + 3)} \\
2 \times 57 = & +\quad 114 \\
\hline
& 41,724
\end{array}
$$

Though you will usually break up the three-digit number when using the addition method, it occasionally pays to break up the two-digit number instead, particularly when the last digit of the two-digit number is a 1 or a 2, as in the following example:

$$
\begin{array}{r}
386 \\
\times \quad 51 \,(50 + 1) \\
\hline
50 \times 386 = \quad 19{,}300 \\
1 \times 386 = +\ \ 386 \\
\hline
19{,}686
\end{array}
$$

This reduces the 3-by-2 to a 3-by-1, made especially easy since the second multiplication problem involves a 1. Note, too, that we were aided in multiplying a 5 by an even number, which produces an extra 0 in the answer, so there is only one digit of overlap in the addition problem.

Another example of multiplying a 5 by an even number is illustrated in the following problem:

$$
\begin{array}{r}
835 \\
\times \quad 62 \,(60 + 2) \\
\hline
60 \times 835 = \quad 50{,}100 \\
2 \times 835 = +\ 1{,}670 \\
\hline
51{,}770
\end{array}
$$

When you multiply the 6 in 60 by the 5 in 835, it generates an extra 0 in the answer, making the addition problem especially easy.

The Subtraction Method

As with 2-by-2s, it is sometimes more convenient to solve a 3-by-2 with subtraction instead of addition, as in the following problems:

$$629 \ (630 - 1)$$
$$\times \quad 38$$

$630 \times 38 = \quad 23{,}940$ $(63 = 9 \times 7)$
$-1 \times 38 = - \quad 38$
$$23{,}902$$

$$758 \ (760 - 2)$$
$$\times \quad 43$$

$760 \times 43 = \quad 32{,}680$ $(43 = 40 + 3)$
$-2 \times 43 = - \quad 86$
$$32{,}594$$

By contrast, you can compare the last subtraction method with the addition method, below, for this same problem:

$$758 \ (750 + 8)$$
$$\times \quad 43$$

$750 \times 43 = \quad 32{,}250$ $(75 = 5 \times 5 \times 3)$
$8 \times 43 = + \quad 344$
$$32{,}594$$

My preference for tackling this problem would be to use the subtraction method because I always try to leave myself with the easiest possible addition or subtraction problem at the end. In this case, I'd rather subtract 86 than add 344, even though the 2-by-2 multiplication problem in the subtraction method above is slightly harder than the one in the addition method.

The subtraction method can also be used for three-digit numbers below a multiple of 100 or close to 1,000, as in the next two examples:

$$293 \ (300 - 7)$$
$$\underline{\times \quad 87}$$

$$300 \times 87 = \quad 26,100$$
$$\underline{-7 \times 87 = - \quad 609}$$
$$25,491$$

$$988 \ (1000 - 12)$$
$$\underline{\times \quad 68}$$

$$1,000 \times 68 = \quad 68,000$$
$$\underline{-12 \times 68 = - \quad 816} \quad (12 = 6 \times 2)$$
$$67,184$$

The last three digits of the answers were obtained by taking the complements of $609 - 100 = 509$ and 816, respectively.

Finally, in the following illustration we break up the two-digit number using the subtraction method. Notice how we subtract 736 by subtracting 1,000 and adding back the complement:

$$736 \qquad\qquad 44,160$$
$$\underline{\times \quad 59} \ (60 - 1) \qquad \underline{- \ 1,000}$$
$$60 \times 736 = \quad 44,160 \qquad\qquad 43,160$$
$$\underline{-1 \times 736 = - \quad 736} \qquad \underline{+ \quad 264} \ \text{(complement of 736)}$$
$$43,424 \qquad\qquad 43,424$$

3-BY-2 EXERCISES USING FACTORING, ADDITION, AND SUBTRACTION METHODS

Solve the 3-by-2 problems below, using the factoring, addition, or subtraction method. Factoring, when possible, is usually easier. The solutions appear in the back of the book.

1.	858 × 15	2.	796 × 19	3.	148 × 62	4.	773 × 42
5.	906 × 46	6.	952 × 26	7.	411 × 93	8.	967 × 51
9.	484 × 75	10.	126 × 87	11.	157 × 33	12.	616 × 37
13.	841 × 72	14.	361 × 41	15.	218 × 68	16.	538 × 53
17.	817 × 61	18.	668 × 63	19.	499 × 25	20.	144 × 56
21.	281 × 44	22.	988 × 22	23.	383 × 49		

The following 3-by-2s will appear in the five-digit squares and the 5-by-5 multiplication problems that follow.

24.	589 × 87	25.	286 × 64	26.	853 × 32	27.	878 × 24
28.	423 × 45	29.	154 × 19	30.	834 × 34	31.	545 × 27
32.	653 × 69	33.	216 × 78	34.	822 × 95		

FIVE-DIGIT SQUARES

Mastering 3-by-2 multiplication takes a fair amount of practice, but once you've managed that, you can slide right into doing five-digit squares because they simplify into a 3-by-2 problem plus a two-digit square and a three-digit square. Watch:

To square the following number:

$$46,792^2$$

Treat it as:

$$46,000 + 792$$
$$\times\ \underline{46,000 + 792}$$

Using the distributive law we can break this down to:

1.   2.   3.
$$46,000 \times 46,000 + 2(46,000)(792) + 792 \times 792$$

This can be stated more simply as:

$$46^2 \times 1\ \text{million} + (46)(792)(2000) + 792^2$$

But I do not do them in this order. In fact, I start in the middle, because the 3-by-2 problems are harder than the two-digit and three-digit squares. So in keeping with the principle of getting the hard stuff out of the way first, I do $792 \times 46 \times 2$ and attach three zeroes on the end, as follows:

$$
\begin{array}{r}
792\ (800 - 8) \\
\times\quad\ \underline{46} \\
800 \times 46 =\quad 36,800 \\
-8 \times 46 = -\quad\underline{368} \qquad \text{"Fisher"}\\
36,432 \times 2,000 = 72,864,000
\end{array}
$$

Using the subtraction method, as shown above, compute $792 \times 46 = 36{,}432$, then double that number to get 72,864. Using the phonetic code from the last chapter on the number 864 allows you to store this number as *72 Fisher*.

The next step is to do $46^2 \times 1$ million, which is 2,116,000,000. At this point, you can say, "Two billion . . ."

Recalling the 72 of *72 Fisher*, you add 116 million to get 188 million. Before saying this number aloud, you need to check ahead to see if there is a carry when adding *Fisher*, or 864, to 792^2. Here you don't actually calculate 792^2; rather, you determine that its product will be large enough to make the 864,000 carry. (You can guesstimate this by noting that 800^2 is 640,000, which will easily make the 864,000 carry, thus you bump the 188 up a notch and say, ". . . 189 million. . . .")

Still holding on to *Fisher*, compute the square of 792, using the three-digit square method (rounding up and down by 8, and so on) to get 627,264. Finally, add 627 to *Fisher*, or 864, to get 1,491. But since you already made the carry, drop the 1 and say, ". . . 491 thousand 264."

Sometimes I forget the last three digits of the answer because my mind has been so preoccupied with the larger computations. So before doing the final addition I will store the 2 of 264 on my fingers and try to remember the 64, which I can usually do because we tend to recall the most recent things heard. If this fails, I can come up with the final two digits by squaring the final two digits of the original number, 92^2, or 8,464, the last two digits of which are the last two digits I'm looking for: 64. (Alternatively, you can convert 264 into a word like *nature*.)

I know this is quite a mouthful. To reiterate the entire problem in a single illustration, here is how I computed $46{,}792^2$:

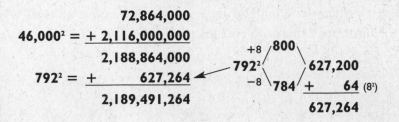

$$792 \ (800 - 8)$$
$$\times \quad 46$$
$$800 \times 46 = \quad 36,800$$
$$-8 \times 46 = - \quad 368 \qquad \text{"Fisher"}$$
$$36,432 \times 2,000 = 72,864,000$$

$$72,864,000$$
$$46,000^2 = + \ 2,116,000,000$$
$$2,188,864,000$$
$$792^2 = + \qquad 627,264$$
$$2,189,491,264$$

$$\begin{array}{c} +8 \diagup 800 \diagdown \\ 792^2 627,200 \\ -8 \diagdown 784 \diagup + \quad 64 \ _{(8^2)} \\ 627,264 \end{array}$$

Let's look at another five-digit square example:

$$83,522^2$$

As before, we compute the answer in this order:

$83 \times 522 \times 2,000, \ 83^2 \times 1$ million, then 522^2

For the first problem, notice that 522 is a multiple of 9. In fact, $522 = 58 \times 9$. Treating 83 as $80 + 3$, we get:

$$522 \ (58 \times 9)$$
$$\times \ 83$$

$$83 \times 58 \times 9 = 4,814 \times 9 = 43,326$$

Doubling 43,326 results in 86,652, which can be stored as 86 *Julian*. Since $83^2 = 6,889$, we can say, "Six billion . . ."

Adding 889 + 86 gives us 975. Before saying 975 million we check to see if *Julian* (652,000) will cause a carry after squaring 522. Guesstimating 522^2 as about 270,000 (500 × 540), you see it will not carry. Thus you can safely say, ". . . 975 million . . ."

Finally, square 522 in the usual way to get 272,484 and add that to *Julian* (652,000) for the rest of the answer: ". . . 924,484."

Illustrated, this problem looks like:

$$83,522^2$$

$$522$$
$$\times\ 83$$

$$83 \times 58 \times 9 = 4,814 \times 9 = 43,326$$

"Julian"

$$
\begin{array}{rl}
43,326 \times 2,000 = & 86,652,000 \\
83,000^2 = + & 6,889,000,000 \\
\hline
& 6,975,652,000 \\
522^2 = + & 272,484 \\
\hline
& 6,975,924,484
\end{array}
$$

$$
\begin{array}{c}
{}^{+22}\diagup 544 \diagdown\ 272,000 \\
522^2 \qquad\qquad + \quad 484 \ (22^2) \\
{}^{-22}\diagdown 500 \diagup\ 272,484
\end{array}
$$

EXERCISE: FIVE-DIGIT SQUARES

1. $45,795^2$ 2. $21,231^2$ 3. $58,324^2$

4. $62,457^2$ 5. $89,854^2$ 6. $76,934^2$

3-BY-3 MULTIPLICATION

In building to the grand finale of 5-by-5 multiplication, 3-by-3s are the final hurdle. As with 3-by-2s, there are a variety of methods you can use to exploit the numbers in the simplification process.

Factoring Method

We'll begin with the factoring method. Unfortunately, most three-digit numbers are not factorable into one-digit numbers, but when they are, the calculation is not too bad.

$$829$$
$$\times\ \underline{288}\ (9 \times 8 \times 4)$$

$$829 \times 9 \times 8 \times 4 = 7,461 \times 8 \times 4 = 59,688 \times 4 = 238,752$$

Notice the sequence involved. You simplify the 3-by-3 (829 × 288) to a 3-by-1-by-1-by-1 through the factoring of 288 into 9 × 8 × 4. This then turns into a 4-by-1-by-1 (7,461 × 8 × 4), and finally into a 5-by-1 to yield the final answer of 238,752. The beauty of this process is that there is nothing to add and nothing to store in memory. When you get to the 5-by-1, you are one step away from completion.

The 5-by-1 problem can be solved in two steps by treating 59,688 as 59,000 + 688, then adding the results of the 2-by-1 (59,000 × 4) and the 3-by-1 (688 × 4), as below:

$$59,688\ (59,000 + 688)$$

$$\times\ \underline{\qquad 4}$$

$$
\begin{array}{rr}
59,000 \times 4 = & 236,000 \\
688 \times 4 = & +\ \underline{\quad 2,752} \\
& 238,752
\end{array}
$$

If both three-digit numbers are factorable into 2-by-1s, then the 3-by-3 can be simplified to a 2-by-2-by-1-by-1, as in the following problem:

$$513 \ (57 \times 9)$$
$$\underline{\times \ 246} \ (41 \times 6)$$

$$57 \times 41 \times 9 \times 6 = 2{,}337 \times 9 \times 6$$
$$= 21{,}033 \times 6$$
$$= 126{,}198$$

As usual, it is best to get the hard part of the problem over first (the 2-by-2). Once you've got this, the problem is then reduced to a 4-by-1, then to a 5-by-1.

More often than not, only one of the three-digit numbers will be factorable, in which case it is reduced to a 3-by-2-by-1, as in the following problem:

$$459 \ (51 \times 9)$$
$$\underline{\times \ 526}$$

$$526 \times 459 = 526 \times 51 \times 9$$
$$= 526 \times (50 + 1) \times 9$$
$$= 26{,}826 \times 9$$
$$= 241{,}434$$

The next 3-by-3 is really just a 3-by-2 in disguise:

$$624$$
$$\underline{\times \ 435}$$

By doubling the 435 and cutting 624 in half, we obtain the equivalent problem:

$$312 \ (52 \times 6)$$
$$\times \ \underline{870} \ (87 \times 10)$$

$$87 \times 52 \times 6 \times 10 = 87 \times (50 + 2) \times 6 \times 10$$
$$= 4{,}524 \times 6 \times 10$$
$$= 27{,}144 \times 10$$
$$= 271{,}440$$

Close-Together Method

Are you ready for something easier? The next multiplication shortcut, which we introduced in Chapter 0, is based on the following algebraic formula:

$$(z + a)(z + b) = z^2 + za + zb + ab$$

Which we rewrite as:

$$(z + a)(z + b) = z(z + a + b) + ab$$

This formula is valid for any values of z, a, and b. We shall take advantage of this whenever the three-digit numbers to be multiplied $(z + a)$ and $(z + b)$ are both near an easy number z (typically one with lots of zeroes in it). For example, to multiply:

$$107$$
$$\times \ \underline{111}$$

We treat this problem as $(100 + 7)(100 + 11)$.
Using $z = 100$, $a = 7$, $b = 11$, our formula gives us:

$$100(100 + 7 + 11) + 7 \times 11 = 100 \times 118 + 77$$
$$= 11{,}877$$

I diagram the problem this way:

$$
\begin{array}{r}
107 \ (7) \\
\times \quad 111 \ (11) \\
\hline
100 \times 118 = \quad 11{,}800 \\
7 \times 11 = + \quad\quad 77 \\
\hline
11{,}877
\end{array}
$$

The numbers in parentheses denote the difference between the number and our convenient "base number" (here, $z = 100$). The number 118 can be obtained either by adding $107 + 11$ or $111 + 7$. Algebraically these sums are always equal since $(z + a) + b = (z + b) + a$.

With fewer words this time, here's another quickie:

$$
\begin{array}{r}
109 \ (9) \\
\times \quad 104 \ (4) \\
\hline
100 \times 113 = \quad 11{,}300 \\
9 \times 4 = + \quad\quad 36 \\
\hline
11{,}336
\end{array}
$$

Neat!

Let's up the ante a little with a higher base number:

$$
\begin{array}{r}
408 \ (8) \\
\times \quad 409 \ (9) \\
\hline
400 \times 417 = \quad 166{,}800 \\
8 \times 9 = + \quad\quad 72 \\
\hline
166{,}872
\end{array}
$$

Although this method is usually used for three-digit multiplication, we can use it for 2-by-2s as well:

$$
\begin{array}{r}
\textbf{78 (8)} \\
\times \quad \textbf{73 (3)} \\
\end{array}
$$

$$
\begin{aligned}
70 \times 81 &= \ \ 5{,}670 \\
8 \times 3 &= +\ \ \ \ \ 24 \\
\hline
&\ \ \ 5{,}694
\end{aligned}
$$

Here, the base number is 70, which we multiply by 81 (78 + 3). Even the addition component is usually very simple.

We can also apply this method when the two numbers are both lower than the base number, as in the following problem where both numbers are just under 400:

$$
\begin{array}{r}
\textbf{396 (−4)} \\
\times \quad \textbf{387 (−13)} \\
\end{array}
$$

$$
\begin{aligned}
400 \times 383 &= \ \ 153{,}200 \\
-4 \times -13 &= +\ \ \ \ \ \ \ 52 \\
\hline
&\ \ 153{,}252
\end{aligned}
$$

The number 383 can be obtained from 396 − 13, or from 387 − 4. I would use this method for 2-by-2 problems like the ones below:

$$
\begin{array}{r}
\textbf{97 (−3)} \\
\times \quad \textbf{94 (−6)} \\
\end{array}
$$

$$
\begin{aligned}
100 \times 91 &= \ \ 9{,}100 \\
-3 \times -6 &= +\ \ \ \ \ 18 \\
\hline
&\ \ \ 9{,}118
\end{aligned}
$$

$$
\begin{array}{r}
\textbf{79 (−1)} \\
\times \quad \textbf{78 (−2)} \\
\end{array}
$$

$$
\begin{aligned}
80 \times 77 &= \ \ 6{,}160 \\
-1 \times -2 &= +\ \ \ \ \ \ 2 \\
\hline
&\ \ \ 6{,}162
\end{aligned}
$$

In our next example, the base number falls between the two numbers:

$$
\begin{array}{r}
396 \,(-4) \\
\times \quad 413 \,(13) \\
\hline
\end{array}
$$

$$400 \times 409 = 163,600$$
$$-4 \times 13 = - \quad\;\; 52$$
$$\overline{\qquad\qquad 163,548}$$

The number 409 is obtained from 396 + 13, or 413 − 4. Notice that since −4 and 13 are of opposite signs we must subtract 52 here.

Let's raise the ante higher still, to where the second step requires a 2-by-2 multiplication:

$$
\begin{array}{r}
621 \,(21) \\
\times \quad 637 \,(37) \\
\hline
\end{array}
$$

$$600 \times 658 = 394,800$$
$$21 \times 37 = + \quad\;\; 777 \;(37 \times 7 \times 3)$$
$$\overline{\qquad\qquad 395,577}$$

We note here that step 1 in the multiplication problem (600 × 658) is itself a reasonable estimate. Our method enables you to go from an estimate to the exact answer.

$$
\begin{array}{r}
876 \,(-24) \\
\times \quad 853 \,(-47) \\
\hline
\end{array}
$$

$$900 \times 829 = 746,100$$
$$-24 \times -47 = + \;\; 1,128 \;(47 \times 6 \times 4)$$
$$\overline{\qquad\qquad 747,228}$$

Also notice that in all these examples, the numbers we multiply in the first step have the same sum as the original numbers.

For example, in the problem above, 900 + 829 = 1729 just as 876 + 853 = 1729. This is because:

$$z + [(z + a) + b] = (z + a) + (z + b)$$

Thus, to obtain the number to be multiplied by 900 (which you know will be 800 plus something), you need only look at the last two digits of 76 + 53 = 129 to determine 829.

In the next problem, adding 827 + 761 = 1588 tells us that we should simply multiply 800 × 788, then *subtract* 27 × 39 as follows:

$$
\begin{array}{r}
827 \ (+27) \\
\times \quad 761 \ (-39) \\
\end{array}
$$

$$
\begin{array}{rl}
800 \times 788 = & 630{,}400 \\
-39 \times 27 = - & 1{,}053 \ (39 \times 9 \times 3) \\
\hline
& 629{,}347
\end{array}
$$

This method is so effective that if the 3-by-3 problem you are presented with has numbers that are not close together, you can sometimes modify the problem by dividing one and multiplying the other, both by the same number, to bring them closer together. For instance, 672 × 157 can be solved by:

$$
\begin{array}{lll}
672 & \div 2 = & 336 \ (36) \\
\times 157 & \times 2 = & \times 314 \ (14) \\
\end{array}
$$

$$
\begin{array}{rl}
300 \times 350 = & 105{,}000 \\
36 \times 14 = + & 504 \ (36 \times 7 \times 2) \\
\hline
& 105{,}504
\end{array}
$$

When the numbers being multiplied are the same (you can't get any closer than that!), notice that the close-together method produces the exact same calculations you did in our traditional squaring procedure:

$$
\begin{array}{r}
347\ (47) \\
\times \quad\ 347\ (47) \\
\hline
300 \times 394 = \quad 118{,}200 \\
47^2 = +\ \ 2{,}209 \\
\hline
120{,}409
\end{array}
$$

Addition Method

When none of the previous methods work, I look for an addition-method possibility, particularly when the first two digits of one of the three-digit numbers is easy to work with. For example, in the problem below, the 64 of 641 is factorable into 8 × 8, so I would solve the problem as illustrated:

$$
\begin{array}{r}
373 \\
\times \quad 641\ (640 + 1) \\
\hline
640 \times 373 = \quad 238{,}720\ \ (373 \times 8 \times 8 \times 10) \\
1 \times 373 = +\ \ \ 373 \\
\hline
239{,}093
\end{array}
$$

In a similar way, in the next problem the 42 of 427 is factorable into 7 × 6, so you can use the addition method and treat 427 as 420 + 7:

$$
\begin{array}{r}
656 \\
\times \quad 427\ (420 + 7) \\
\hline
420 \times 656 = \quad 275{,}520\ \ (656 \times 7 \times 6 \times 10) \\
7 \times 656 = +\ \ 4{,}592 \\
\hline
280{,}112
\end{array}
$$

Often I break the last addition problem into two steps, as follows:

$$
\begin{array}{r}
275,520 \\
7 \times 600 = {} + \underline{4,200} \\
279,720 \\
7 \times 56 = {} + \underline{392} \\
280,112
\end{array}
$$

Since addition-method problems can be very strenuous, I usually go out of my way to find a method that will produce a simple computation at the end. For example, the above problem could have been done using the factoring method. In fact, that is how I would choose to do it:

$$
\begin{array}{r}
656 \\
\underline{\times\ 427}\ (61 \times 7)
\end{array}
$$

$$
\begin{aligned}
656 \times 61 \times 7 &= 656 \times (60 + 1) \times 7 \\
&= 40,016 \times 7 \\
&= 280,112
\end{aligned}
$$

The simplest addition-method problems are those in which one number has a 0 in the middle, as below:

$$
\begin{array}{r}
732 \\
\underline{\times\ \ \ \ \ 308}\ (300 + 8) \\
300 \times 732 = {}\ \ \ 219,600 \\
8 \times 732 = {} + \underline{\ \ \ 5,856} \\
225,456
\end{array}
$$

These problems tend to be so much easier than other addition-method problems that it pays to see whether the 3-by-3 can be converted to a problem like this. For instance, 732 × 308 could have been obtained by either of the "non-zero" problems below:

$$244 \times 3 = \quad 732 \qquad \text{or} \qquad 366 \times 2 = \quad 732$$
$$\underline{\times\, 924 \div 3} = \underline{\times\, 308} \qquad\qquad \underline{\times\, 616 \div 2} = \underline{\times\, 308}$$

We mention that another way to do this problem is by 308 × 366 × 2, and take advantage of the closeness of 308 and 366. Let's do one more toughie:

$$
\begin{array}{rr}
 & 739 \\
 \times & \underline{443} \ (440 + 3) \\
440 \times 739 = & 325{,}160 \ (739 \times 11 \times 4 \times 10) \\
3 \times 700 = & \underline{+\quad 2{,}100} \\
 & 327{,}260 \\
3 \times 39 = & \underline{+\qquad 117} \\
 & 327{,}377
\end{array}
$$

Subtraction Method

The subtraction method is one that I sometimes use when one of the three-digit numbers can be rounded up to a convenient two-digit number with a 0 at the end, as in the next problem:

$$
\begin{array}{rr}
 & 719 \ (720 - 1) \\
 \times & \underline{247} \\
720 \times 247 = & 177{,}840 \ (247 \times 9 \times 8 \times 10) \\
-1 \times 247 = & \underline{-\quad 247} \\
 & 177{,}593
\end{array}
$$

Likewise in the following example:

$$
\begin{array}{r}
538\ (540 - 2) \\
\times\quad 346 \\
\end{array}
$$

$$540 \times 346 =\ 186{,}840\ \text{(346} \times \text{6} \times \text{9} \times \text{10)}$$
$$-2 \times 346 = -\quad 692$$
$$\overline{\qquad\ \ 186{,}148}$$

When-All-Else-Fails Method

When all else fails, I use the following method, which is fool-proof when you can find no other method to exploit the numbers. In the when-all-else-fails method, the 3-by-3 problem is broken down into three parts: a 3-by-1, a 2-by-1, and a 2-by-2. As you do each computation, you sum the totals as you go. These problems are difficult, especially if you cannot see the original number. In my presentation of 3-by-3s and 5-by-5s, I have the problems written down, but I do all the calculations mentally.

Here's an example:

$$
\begin{array}{r}
851 \\
\times\quad 527 \\
\end{array}
$$

$$500 \times 851 =\quad 425{,}500$$
$$27 \times 800 = +\ 21{,}600$$
$$\overline{\qquad\ \ 447{,}100}$$
$$27 \times 51 = +\quad 1{,}377$$
$$\overline{\qquad\ \ 448{,}477}$$

In practice, the calculation actually proceeds as shown below. Sometimes I use the phonetic code to store the thousands digits (e.g., 447 = *our rug*) and put the hundreds digit (1) on my fingers:

$$851$$
$$\times\ 527$$
$$5 \times 851 =\ \ 4{,}255$$
$$8 \times 27 = +\ 216 \qquad \text{"Our rug"}$$
$$4{,}471 \times 100 = 447{,}100$$
$$51 \times 27 = +\ 1{,}377$$
$$448{,}477$$

Let's do another example, but this time I'll break up the first number. (I usually break up the larger one so that the addition problem is easier.)

$$923$$
$$\times\ 673$$
$$9 \times 673 =\ \ 6{,}057$$
$$6 \times 23 = +\ 138 \qquad \text{"Shut up"}$$
$$6{,}195 \times 100 = 619{,}500$$
$$73 \times 23 = +\ 1{,}679$$
$$621{,}179$$

EXERCISE: 3-BY-3 MULTIPLICATION

1.	644	2.	596	3.	853	4.	343	5.	809
	× 286		× 167		× 325		× 226		× 527

6.	942	7.	692	8.	446	9.	658	10.	273
	× 879		× 644		× 176		× 468		× 138

11.   **824**   12.   **642**   13.   **783**   14.   **871**   15.   **341**
× 206   **× 249**   **× 589**   **× 926**   **× 715**

16.   **417**   17.   **557**   18.   **976**   19.   **765**
× 298   **× 756**   **× 878**   **× 350**

The following problems are embedded in the 5-by-5 multiplication problems in the next section:

20.   **154**   21.   **545**   22.   **216**   23.   **393**
× 423   **× 834**   **× 653**   **× 822**

5-BY-5 MULTIPLICATION

The largest problem we will attempt is the mental multiplication of two five-digit numbers. To do a 5-by-5, you need to have mastered 2-by-2s, 2-by-3s, and 3-by-3s, as well as the phonetic code. Now it is just a matter of putting it all together. As you did in the five-digit square problems, you will use the distributive law to break down the numbers. For example:

$$\text{27,639 (27,000 + 639)}$$
$$\underline{\times \text{ 52,196 (52,000 + 196)}}$$

Based on this you can break down the problem into four easier multiplication problems, which I illustrate below in crisscross fashion as a 2-by-2, two 3-by-2s, and finally a 3-by-3, summing the results for a grand total. That is,

$$\text{(27 × 52) million}$$
$$+ \text{ [(27 × 196) + (52 × 639)] thousand}$$
$$+ \text{ (639 × 196)}$$

As with the five-digit squares, I begin in the middle with the 3-by-2s, starting with the harder 3-by-2:

"Mom, no knife"

1. $52 \times 639 = 52 \times 71 \times 9 = 3{,}692 \times 9 = 33{,}228$

Committing 33,228 to memory with the mnemonic *Mom, no knife,* you then turn to the second 3-by-2:

2. $27 \times 196 = 27 \times (200 - 4) = 5{,}400 - 108 = 5{,}292$

and add it to the number that you are storing:

3.   33,228 ("Mom, no knife")
     $\underline{+ \ 5{,}292}$
     38,520

for a new total, which we store as:

"Movie lines" (38 million, 520 thousand)

Holding on to *movie lines,* compute the 2-by-2:

4. $52 \times 27 = 52 \times 9 \times 3 = 1{,}404$

At this point, you can give part of the answer. Since this 2-by-2 represents 52×27 million, 1,404 represents 1 billion, 404 million. Since 404 million will not cause a carry, you can safely say, "One billion . . ."

5. $404 +$ "Movie" $(38) = 442$

In this step you add 404 to *movie* (38) to get 442, at which point you can say, "...442 million...." You can say this because you know 442 will not carry—you've peeked ahead at the 3-by-3 to see whether it will cause 442 to carry to a higher number. If you found that it would carry, you would say, "443 million." But since *lines* is 520,000 and the 3-by-3 (639 × 196) will not exceed 500,000 (a rough guesstimate of 600 × 200 = 120,000 shows this), you are safe.

6. **639 × 196 = 639 × 7 × 7 × 4 = 4,473 × 7 × 4**
    = 31,311 × 4 = 125,244

While still holding on to *lines,* you now compute the 3-by-3 using the factoring method, to get 125,244. You might convert 244 into a word like *nearer.* The final step is a simple addition of:

7. **125,244 + "Lines" (520,000)**

This allows you to say the rest of the answer: "...645,244." Since a picture is worth a thousand calculations, here's our picture of how this would look:

27,639
× 52,196

"Mom, no knife"

639 × 52 =   33,228
196 × 27 = + 5,292   "Movie lines"

38,520 × 1,000 =   38,520,000
52 × 27 × 1 million = + 1,404,000,000
1,442,520,000
639 × 196 = +   125,244
1,442,645,244

I should make a parenthetical note here that I am assuming in doing 5-by-5s that you can write down the problem on a blackboard or piece of paper. If you can't, you will have to create a mnemonic for each of the four numbers. For example, in the last problem, you could store the problem as:

$$27,639\text{—"Neck jump"}$$
$$\times\ 52,196\text{—"Lion dopish"}$$

Then you would multiply *lion* × *jump, dopish* × *neck, lion* × *neck,* and finally *dopish* × *jump*. Obviously this would slow you down a bit, but if you want the extra challenge of not being able to see the numbers, you can still solve the problem.

We conclude with one more 5-by-5 multiplication:

$$79,838$$
$$\times\ 45,547$$

The steps are the same as those in the last problem. You start with the harder 3-by-2 and store the answer with a mnemonic:

1. $547 \times 79 = 547 \times (80 - 1) = 43,760 - 547$
$= 43,213$—"Rome anatomy"

Then you compute the other 3-by-2.

2. $838 \times 45 = 838 \times 5 \times 9 = 4,190 \times 9 = 37,710$

Summing the 3-by-2s you commit the new total to memory.

3.   . **43,213**—"Rome anatomy"
   + 37,710
   80,923—"Face Panama"

4.   79 × 45 = 79 × 9 × 5 = 711 × 5 = 3,555

The 2-by-2 gives you the first digit of the final answer, which you can say out loud with confidence: "Three billion . . ."

5.   555 + "Face" **(80) = 635**

The millions digits of the answer involve a carry from 635 to 636, because *Panama* (923) needs only 77,000 to cause it to carry, and the 3-by-3 (838 × 547) will easily exceed that figure. So you say: ". . . 636 million . . ."

The 3-by-3 is computed using the addition method:

6.   **838**
                          ×   547 (540 + 7)
        540 × 838 =   452,520 (838 × 9 × 6 × 10)
        7 × 800 = +   5,600
                          458,120
        7 × 38 = +   266
                          458,386

And in the next step you add this total to *Panama* (923,000):

7.   923,000
     + 458,386
     1,381,386

Since you've already used the 1 in the carry to 636 million, you just say the thousands: ". . . 381 thousand . . . 386," and take a bow!

This problem may be illustrated in the following way:

79,838
× 45,547

"Rome anatomy"

547 × 79 =   43,213
838 × 45 = +37,710   "Face Panama"

                    80,923 × 1,000 =   80,923,000
              79 × 45 × 1 million = + 3,555,000,000
                                                3,635,923,000
                    838 × 547 = +   458,386
                                                3,636,381,386

EXERCISE: 5-BY-5 MULTIPLICATION

1.   65,154   2.   34,545   3.   69,216   4.   95,393
   × 19,423   × 27,834   × 78,653   × 81,822

Chapter 9

Presto-digit-ation: The Art of Mathematical Magic

Playing with numbers has brought me great joy in life. I find that arithmetic can be just as entertaining as magic. But understanding the magical secrets of arithmetic requires algebra. Of course, there are other reasons to learn algebra (SATs, modeling real-world problems, computer programming, and understanding higher mathematics, to name just a few), but what first got me interested in algebra was the desire to understand some mathematical magic tricks, which I now present to you!

PSYCHIC MATH

Say to a volunteer in the audience, "Think of a number, any number." And you should also say, "But to make it easy on yourself, think of a one-digit or two-digit number." After you've reminded your volunteer that there's no way you could know her number, ask her to:

1. double the number,
2. add 12,

3. divide the total by 2,
4. subtract the original number.

Then say, "Are you now thinking of the number six?" Try this one on yourself first and you will see that the sequence always produces the number 6 no matter what number is originally selected.

Why This Trick Works

This trick is based on simple algebra. In fact, I sometimes use this as a way to introduce algebra to students. The secret number that your volunteer chose can be represented by the letter x. Here are the functions you performed in the order you performed them:

1. $2x$ (double the number)
2. $2x + 12$ (then add 12)
3. $(2x + 12) \div 2 = x + 6$ (then divide by 2)
4. $x + 6 - x = 6$ (then subtract original number)

So no matter what number your volunteer chooses, the final answer will always be 6. If you repeat this trick, have the volunteer add a different number at step 2 (say 18). The final answer will be half that number (namely 9).

THE MAGIC 1089!

Here is a trick that has been around for centuries. Have your audience member take out paper and pencil and:

1. secretly write down a three-digit number where the digits are decreasing (like 851 or 973),

2. reverse that number and subtract it from the first number,
3. take that answer and add it to the reverse of itself.

At the end of this sequence, the answer 1089 will magically appear, no matter what number your volunteer originally chose. For example:

$$\begin{array}{r} 851 \\ -\ 158 \\ \hline 693 \\ +\ 396 \\ \hline 1089 \end{array}$$

Why This Trick Works

No matter what three-digit number you or anyone else chooses in this game, the final result will always be 1089. Why? Let *abc* denote the unknown three-digit number. Algebraically, this is equal to:

$$100a + 10b + c$$

When you reverse the number and subtract it from the original number, you get the number *cba*, algebraically equal to:

$$100c + 10b + a$$

Upon subtracting *cba* from *abc*, you get:

$$100a + 10b + c - (100c + 10b + a)$$
$$= 100(a - c) + (c - a)$$
$$= 99(a - c)$$

Hence, after subtracting in step 2, we must have one of the following multiples of 99: 198, 297, 396, 495, 594, 693, 792, or 891, each one of which will produce 1089 after adding it to the reverse of itself, as we did in step 3.

MISSING-DIGIT TRICKS

Using the number 1,089 from the last effect, hand a volunteer a calculator and ask her to multiply 1089 by any three-digit number she likes, but not to tell you the three-digit number. (Say she secretly multiplies 1,089 × 256 = 278,784.) Ask her how many digits are in her answer. She'll reply, "Six."

Next you say: "Call out five of your six digits to me in any order you like. I shall try to determine the missing digit."

Suppose she calls out, "Two . . . four . . . seven . . . eight . . . eight." You correctly tell her that she left out the number 7.

The secret is based on the fact that a number is a multiple of 9 if and only if its digits sum to a multiple of 9. Since $1 + 0 + 8 + 9 = 18$ is a multiple of 9, then so is 1089. Thus 1089 times any whole number will also be a multiple of 9. Since the digits called out add up to 29, and the next multiple of 9 greater than 29 is 36, our volunteer must have left out the number 7 (since $29 + 7 = 36$).

There are more subtle ways to force the volunteer to end up with a multiple of 9. Here are some of my favorites:

1. Have the volunteer randomly choose a six-digit number, scramble its digits, then subtract the smaller six-digit number from the larger one. Since we're subtracting two numbers with the same mod sum (indeed, the identical sum of digits), the resulting difference will have a mod sum of 0, and hence be a multiple of 9. Then continue as before to find the missing digit.

2. Have the volunteer secretly choose a four-digit number, reverse the digits, then subtract the smaller number from the larger. (This will be a multiple of 9.) Then multiply this by any three-digit number, and continue as before.

3. Ask the volunteer to multiply one-digit numbers randomly until the product is seven digits long. This is not "guaranteed" to produce a multiple of 9, but in practice it will do so at least 90% of the time (the chances are high that the one-digit numbers being multiplied include a 9 or two 3s or two 6s, or a 3 and a 6). I often use this method in front of mathematically advanced audiences who might see through other methods.

There is one problem to watch out for. Suppose the numbers called out add up to a multiple of 9 (say 18). Then you have no way of determining whether the missing digit is 0 or 9. How do you remedy that? Simple—you cheat! You merely say, "You didn't leave out a zero, did you?" If she did leave out a 0, you have completed the trick successfully. If she did not leave out the 0, you say: "Oh, it seemed as though you were thinking of nothing! You didn't leave out a one, two, three, or four, did you?" She'll either shake her head, or say no. Then you follow with, "Nor did you leave out a five, six, seven, or eight, either. You left out the number nine, didn't you?" She'll respond in the affirmative, and you will receive your well-deserved applause!

LEAPFROG ADDITION

This trick combines a quick mental calculation with an astonishing prediction. Handing the spectator a card with ten lines, numbered 1 through 10, have the spectator think of two positive numbers between 1 and 20, and enter them on lines 1 and 2 of the card. Next have the spectator write the sum of lines 1 and

2 on line 3, then the sum of lines 2 and 3 on line 4, and so on as illustrated below.

1	9
2	2
3	11
4	13
5	24
6	37
7	61
8	98
9	159
10	257

Finally, have the spectator show you the card. At a glance, you can tell him the sum of all the numbers on the card. For instance, in our example, you could instantly announce that the numbers sum up to 671 faster than the spectator could do using a calculator! As a kicker, hand the spectator a calculator, and ask him to divide the number on line 10 by the number on line 9. In our example, the quotient $\frac{257}{159} = 1.616\ldots$. Have the spectator announce the first three digits of the quotient, then turn the card over (where you have already written a prediction). He'll be surprised to see that you've already written the number 1.61!

Why This Trick Works

To perform the quick calculation, you simply multiply the number on line 7 by 11. Here $61 \times 11 = 671$. The reason this works is illustrated in the table below. If we denote the numbers on lines 1 and 2 by x and y, respectively, then the sum of lines 1 through 10 must be $55x + 88y$, which equals 11 times $(5x + 8y)$, that is, eleven times the number on line 7.

1	x
2	y
3	$x + y$
4	$x + 2y$
5	$2x + 3y$
6	$3x + 5y$
7	$5x + 8y$
8	$8x + 13y$
9	$13x + 21y$
10	$21x + 34y$
Total:	$55x + 88y$

As for the prediction, we exploit the fact that for any positive numbers, $a, b, c, d,$ if $a/b < c/d,$ then it can be shown that the fraction you get by "adding fractions badly" (i.e., adding the numerators and adding the denominators) must lie in between the original two fractions. That is,

$$\frac{a}{b} < \frac{a+c}{b+d} < \frac{c}{d}$$

Thus the quotient of line 10 divided by line 9 $(21x + 34y)/(13x + 21y)$, must lie between

$$1.615\ldots = \frac{21x}{13x} < \frac{21x + 34y}{13x + 21y} < \frac{34y}{21y} = 1.619\ldots$$

Hence, the ratio must begin with the digits 1.61, as predicted.

In fact, if you continue the leapfrog process indefinitely, the ratio of consecutive terms gets closer and closer to

$$\frac{1 + \sqrt{5}}{2} \approx 1.6180339887\ldots$$

a number with so many amazingly beautiful and mysterious properties that it is often called the golden ratio.

MAGIC SQUARES

Are you ready for a challenge of a different sort? Below you will find what is called a magic square. There has been much written on magic squares and how to construct them, going back as far as ancient China. Here we describe a way to present magic squares in an entertaining fashion. This is a routine I've been doing for years.

I bring out a business card with the following written on the back:

8	11	14	1
13	2	7	12
3	16	9	6
10	5	4	15

= 34

I say, "This is called a magic square. In fact, it's the smallest magic square you can create, using the numbers one through sixteen. You'll notice that every row and every column adds to the same number—thirty-four. Now I've done such an extensive study on magic squares that I propose to create one for you right before your very eyes."

I then ask someone from the audience to give me any number larger than 34. Let's suppose she says 67.

I then bring out another business card and draw a blank 4-by-4 grid, and place the number 67 to the right of it. Next I ask her to point to the squares, one at a time, in any order. As she points to an empty box, I immediately write a number inside it. The end result looks like this:

16	19	23	9
22	10	15	20
11	25	17	14
18	13	12	24

= 67

I continue: "Now with the first magic square, every row and column added to thirty-four. [I usually put the thirty-four card away at this point.] Let's see how we did with your square." After checking that each row and column adds up to 67, I say: "But I did not stop there. For you, I decided to go one step further. Notice that both *diagonals* add up to sixty-seven!" Then I point out that the four squares in the *upper left corner* sum to 67 (16 + 19 + 22 + 10 = 67), as do the other three four-square corners, the four squares *in the middle,* and the four *corner squares*! "They all sum to sixty-seven. But don't take my word for any of this. Please keep this magic square as a souvenir from me—and check it out for yourself!"

HOW TO CONSTRUCT A MAGIC SQUARE

You can create a magic square that sums to any number by taking advantage of the original magic square that sums to 34. Keep that square within eyeshot while you construct the volunteer's magic square. As you draw the 4-by-4 grid, mentally perform the calculations of steps 1 and 2:

1. Subtract 34 from the given number (e.g., 67 − 34 = 33).
2. Divide this number by 4 (e.g., 33 ÷ 4 = 8 with a remainder of 1).

The quotient is the first "magic" number. The quotient plus the remainder is the second "magic" number. (Here our magic numbers are 8 and 9.)

3. When the volunteer points out a square, inconspicuously look at the 34-square and see what is in the corresponding square. If it is a 13, 14, 15, or 16, add the second number to it (e.g., 9). If not, add the first magic number (e.g., 8).

4. Insert the appropriate number until the magic square is completed.

Note that when the given number is even, but not a multiple of 4, then your first and second magic numbers will be the same, so you'll have just one magic number to add to the numbers in your 34-square.

Why This Trick Works

The reason this method works is based on the fact that every row, column, diagonal (and more) from the originally displayed magic square sums to 34. Suppose the given number had been 82. Since 82 − 34 = 48 (and 48 ÷ 4 = 12), we would add 12 to each square. Then every group of four that had previously summed to 34 would add to 34 + 48 = 82. See the magic square below.

20	23	26	13
25	14	19	24
15	28	21	18
22	17	16	27

= 82

On the other hand, if the given number were 85, our magic numbers would be 12 and 15, so we would be adding 3 more to the squares showing 13, 14, 15, and 16. Since each row, column, and group of four contains exactly one of these numbers, each group of four would now add to $34 + 48 + 3 = 85$ in the following magic square.

20	23	29	13
28	14	19	24
15	31	21	18
22	17	16	30

$= 85$

As an interesting piece of mathemagical trivia, let me point out another astonishing property of the famous 3-by-3 magic square below.

4	9	2
3	5	7
8	1	6

$= 15$

Not only do the rows, columns, and diagonals add up to 15, but if you treat the rows of the magic square as three-digit numbers, you can verify on your calculator that $492^2 + 357^2 + 816^2 = 294^2 + 753^2 + 618^2$. Also, $438^2 + 951^2 + 276^2 = 834^2 + 159^2 + 672^2$. If you are curious about *why* this property happens, you might want to explore my paper *Magic "Squares" Indeed!* (included in the bibliography).

QUICK CUBE ROOTS

Ask someone to select a two-digit number and keep it secret. Then have him cube the number; that is, multiply it by itself twice (using a calculator). For instance, if the secret number is 68, have the volunteer compute $68 \times 68 \times 68 = 314,432$. Then ask the volunteer to tell you his answer. Once he tells you the cube, 314,432, you can instantly reveal the original secret number, the cube root, 68. How?

To calculate cube roots, you need to learn the cubes from 1 to 10:

$$1^3 = 1$$
$$2^3 = 8$$
$$3^3 = 27$$
$$4^3 = 64$$
$$5^3 = 125$$
$$6^3 = 216$$
$$7^3 = 343$$
$$8^3 = 512$$
$$9^3 = 729$$
$$10^3 = 1000$$

Once you have learned these, calculating the cube roots is as easy as π. For instance, with this example problem:

What is the cube root of 314,432?

Seems like a pretty tough one to begin with, but don't panic, it's actually quite simple. As usual, we'll take it one step at a time:

1. Look at the magnitude of the thousands number (the numbers to the left of the comma), 314 in this example.
2. Since 314 lies between $6^3 = 216$ and $7^3 = 343$, the cube root lies in the 60s (since $60^3 = 216,000$ and $70^3 = 343,000$). Hence the first digit of the cube root is 6.
3. To determine the last digit of the cube root, note that only the number 8 has a cube that ends in 2 ($8^3 = 512$), so the last digit ends in 8.

Therefore, the cube root of 314,432 is 68. Three simple steps and you're there. Notice that every digit, 0 through 9, appears once among the last digits of the cubes. (In fact, the last digit of the cube root is equal to the last digit of the cube of the last digit of the cube. Go figure out that one!)

Now you try one for practice:

What is the cube root of 19,683?
1. 19 lies between 8 and 27 (2^3 and 3^3).
2. Therefore the cube root is 20-something.
3. The last digit of the problem is 3, which corresponds to 343 = 7^3, so 7 is the last digit.

The answer is 27.

Notice that our derivation of the last digit will only work if the original number is the cube root of a whole number. For instance, the cube root of 19,684 is 27.0004572 . . . definitely not 24. That's why we included this in our mathemagical magic section and not in an earlier chapter. (Besides, the calculation goes so fast, it seems like magic!)

SIMPLIFIED SQUARE ROOTS

Square roots can also be calculated easily if you are given a perfect square. For instance, if someone told you that the square of a two-digit number was 7569, you could immediately tell her that the original number (the square root) is 87. Here's how.

1. Look at the magnitude of the "hundreds number" (the numbers preceding the last two digits), 75 in this example.
2. Since 75 lies between 8^2 (8 × 8 = 64) and 9^2 (9 × 9 = 81), then we know that the square root lies in the 80s. Hence the first digit of

the square root is 8. Now there are two numbers whose square ends in 9: $3^2 = 9$ and $7^2 = 49$. So the last digit must be 3 or 7. Hence the square root is either 83 or 87. Which one?

3. Compare the original number with the square of 85 (which we can easily compute as $80 \times 90 + 25 = 7225$. Since 7569 is larger than 7225, the square root is the larger number, 87.

Let's do one more example.

What is the square root of 4761?

Since 47 lies between $6^2 = 36$ and $7^2 = 49$, the answer must be in the 60s. Since the last digit of the square is 1, the last digit of the square root must be 1 or 9. Since 4761 is greater than $65^2 = 4225$, the square root must be 69. As with the previous cube root trick, this method can be applied only when the original number given is a perfect square.

AN "AMAZING" SUM

The following trick was first shown to me by James "the Amazing" Randi, who has used it effectively in his magic. Here, the magician is able to predict the total of four randomly chosen three-digit numbers.

To prepare this trick you will need three sets of nine cards each, and a piece of paper with the number 2247 written down on it and then sealed in an envelope. Next, on each of the three sets of cards do the following:

On Set A write the following numbers, one number on each card:

4286   5771   9083   6518   2396   6860   2909   5546   8174

On Set B write the following numbers:

5792   6881   7547   3299   7187   6557   7097   5288   6548

On Set C write the following numbers:

2708   5435   6812   7343   1286   5237   6470   8234   5129

Select three people in the audience and give each one a set of cards. Have each of your volunteers randomly pick one of the nine cards they hold. Let's say they choose the numbers 4286, 5792, and 5435. Now, in the sequence, have each one call out one digit from the four-digit number, first person A, then person B, and finally person C. Say they call out the numbers 8, 9, and 5. Write down the numbers 8, 9, and 5 (895) and say, "You must admit that this number was picked entirely at random and could not possibly have been predicted in advance."

Next, have the three people call out a different number from their cards. Say they call out 4, 5, and 3. Write 453 below 895. Then repeat this two more times with their remaining two numbers, resulting in four three-digit numbers, such as:

	A	B	C
	8	9	5
	4	5	3
	2	2	4
	6	7	5
2	2	4	7

Next have someone add the four numbers and announce the total. Then have someone open the envelope to reveal your prediction. Take a bow!

Why This Trick Works

Look at the numbers in each set of cards and see if you can find anything consistent about them. Each set of numbers sums to the same total. Set A numbers total to 20. Set B numbers total to 23. Set C numbers total to 17. With person C's numbers in the right column totaling to 17, you will always put down the 7 and carry the 1. With person B's numbers totaling to 23, plus the 1, you will always put down the 4 and carry the 2. Finally with person A's numbers totaling to 20, adding the 2 gives you a total of 2247!

A DAY FOR ANY DATE

We conclude our book with one of the classic feats of mental calculation: how to figure out the day of the week of anyone's birthday. This is actually a very practical skill. It's not every day that someone will ask you to square a three-digit number, but hardly a day goes by without somebody mentioning a date to you in the past or future. With just a little bit of practice, you can quickly and easily determine the day of the week of practically any date in history.

First we assign a code number to every day of the week. They are easy to remember:

Number	Day
1	Monday
2	Tuesday
3	Wednesday
4	Thursday
5	Friday
6	Saturday
7 or 0	Sunday

I have found the above list to be easy to remember, especially since 2's day is Tuesday. The other days can be given similar mnemonics: "1 *day* is Monday, 4's day is Thursday, and 5 *day* is Friday. For Wednesday, notice that if you hold up three fingers, you get the letter *w*. For the weekends (at the end of our list), you might think of Saturday as 6-*urday* and Sunday as 7 *day* (especially if you pronounce the *v* softly). Or to remember the zero, you could think of Sunday as *none-day* or *nun-day*!

Next we need a code for every month of the year. These month codes are used for every year, with two exceptions. In a leap year (like 2000 or 2004 or 2008 or . . .) the month code for January is 5, and the month code for February is 1. To make the month codes easier to remember, we provide a table of mnemonics below.

Month	Code	Mnemonic
January	6*	W-I-N-T-E-R has 6 letters.
February	2*	February is the 2nd month of the year.
March	2	March 2 the beat of the drum!
April	5	A-P-R-I-L and F-O-O-L-S have 5 letters.
May	0	May I have a sandwich? Hold the May-0!
June	3	June B-U-G has 3 letters.
July	5	Watching FIVER-works and FIVER-crackers!
August	1	August begins with A, the 1st letter.
September	4	September is the beginning of F-A-L-L.
October	6	Halloween T-R-I-C-K-S and T-R-E-A-T-S.
November	2	I'll have 2 servings of TUrkey, please!
December	4	December is the L-A-S-T month, or X-M-A-S.

*In a leap year, the code for January is 5 and the code for February is 1.

Now let's calculate the day of the week for any date in 2006. After that, we will describe 2007, then 2008, and so on, for the rest of your life. Once all future dates are taken care of, we can look back into the past and determine the days of the week for any date in the 1900s or any other century.

Every year is assigned a code number, and for 2006 that year code happens to be 0 (see page 218).

Now, to calculate the day of the week, you simply add the month code plus the date code plus the year code. Thus for December 3, 2006, we compute

Month Code + Date + Year Code = 4 + 3 + 0 = 7

Hence, this date will be on Day 7, which is Sunday.

How about November 18, 2006? Since November has a month code of 2, we have

Month Code + Date + Year Code = 2 + 18 + 0 = 20.

Now since the week repeats every seven days, we can subtract any multiple of 7 from our answer (7, 14, 21, 28, 35, . . .) and this will not change the day of the week. So our final step is to subtract the biggest multiple of 7 to get 20 − 14 = 6. Hence November 18, 2006, occurs on Saturday.

How about 2007? Well, what happens to your birthday as you go from one year to the next? For most years, there are 365 days, and since 364 is a multiple of 7 (7 × 52 = 364), then the day of the week of your birthday will shift forward by one day in most years. If there are 366 days between your birthdays, then it will shift forward by two days. Hence, for 2007 we calculate the day of the week just as before, but now we use a year code of 1. Next, 2008 is a leap year. (Leap years occur every four years, so 2000, 2004, 2008, 2012, . . . , 2096 are the leap years of the twenty-first century.) Hence, for 2008, the year code increases by two, so it will have a year code of 3. The next year, 2009, is not a leap year, so the year code increases by one, to 4.

Thus, for example, May 2, 2007, has

Month Code + Date + Year Code = 0 + 2 + 1 = 3

so this date is a Wednesday.

For September 9, 2008, we have

Month Code + Date + Year Code = 4 + 9 + 3 = 16

Subtracting the biggest multiple of 7, we have $16 - 14 = 2$, so this date is a Tuesday.

On the other hand, January 16, 2008, is a leap year January, so its month code will be 5 instead of 6. Thus, we have

Month Code + Date + Year Code = 5 + 16 + 3 = 24

and therefore occurs on day $24 - 21 = 3$, which is Wednesday. For your reference, we have listed all of the year codes for the twenty-first century in the figure on the following page.

The good news is that we do not have to memorize this table. We can mentally calculate the year code for any date between 2000 and 2099. For the year code of $2000 + x$, we simply take the number $x/4$ (ignoring any remainder) and add that to x. The year code can be reduced by subtracting any multiple of 7.

For example, with 2061, we see that $\frac{61}{4} = 15$ (with a remainder of 1 that we ignore). Thus 2061 has a year code of $61 + 15 = 76$. And since we can subtract any multiple of 7, we use the simpler year code of $76 - 70 = 6$.

Year	Code	Year	Code	Year	Code	Year	Code
2000	0	2025	3	2050	6	2075	2
2001	1	2026	4	2051	0	2076	4
2002	2	2027	5	2052	2	2077	5
2003	3	2028	0	2053	3	2078	6
2004	5	2029	1	2054	4	2079	0
2005	6	2030	2	2055	5	2080	2
2006	0	2031	3	2056	0	2081	3
2007	1	2032	5	2057	1	2082	4
2008	3	2033	6	2058	2	2083	5
2009	4	2034	0	2059	3	2084	0
2010	5	2035	1	2060	5	2085	1
2011	6	2036	3	2061	6	2086	2
2012	1	2037	4	2062	0	2087	3
2013	2	2038	5	2063	1	2088	5
2014	3	2039	6	2064	3	2089	6
2015	4	2040	1	2065	4	2090	0
2016	6	2041	2	2066	5	2091	1
2017	0	2042	3	2067	6	2092	3
2018	1	2043	4	2068	1	2093	4
2019	2	2044	6	2069	2	2094	5
2020	4	2045	0	2070	3	2095	6
2021	5	2046	1	2071	4	2096	1
2022	6	2047	2	2072	6	2097	2
2023	0	2048	4	2073	0	2098	3
2024	2	2049	5	2074	1	2099	4

Hence March 19, 2061, has

Month Code + Date + Year Code = 2 + 19 + 6 = 27

Subtracting $27 - 21 = 6$ tells us that this date will take place on Saturday.

What about birth dates between 1900 and 1999? Do the problem exactly as in the previous calculation, but shift the final answer forward by one day (or simply add 1 to the year code). Thus March 19, 1961, occurred on a Sunday.

For the date December 3, 1998, we have $\frac{98}{4} = 24$ (with a remainder of 2 that we ignore). Thus 1998 has year code $98 + 24 + 1 = 123$, where the plus one is added for dates in the 1900s. Next, subtract the biggest multiple of 7. For handy reference, here are the multiples of 7 that you might need:

7, 14, 21, 28, 35, 42, 49, 56, 63, 70, 77, 84, 91, 98, 105, 112, 119, 126, . . .

Since $123 - 119 = 4$, 1998 has a year code of 4. Therefore, December 3, 1998, has

Month Code + Date + Year Code = 4 + 3 + 4 = 11

and $11 - 7 = 4$, so this date occurred on a Thursday.

For dates in the 1800s, we add 3 to the year code. For example, Charles Darwin and Abraham Lincoln were both born on February 12, 1809. Since 2009 has a year code of 4, then 1809 has a year code of $4 + 3 = 7$, which can be reduced to 0. Thus, February 12, 1809, has

Month Code + Date + Year Code = 2 + 12 + 0 = 14

and $14 - 14 = 0$, so they were both born on a Sunday.

For dates in the 2100s, we add 5 to the year code (or, equivalently, subtract 2 from the year code). For example, since 2009 has a year code of 4, then 2109 has a year code of 4 + 5 = 9, which, after subtracting 7 is the same as a year code of 2. Dates in the 1700s are treated just like in the 2100s (by adding 5 or subtracting 2) but we need to be careful. The days that we are calculating are based on the Gregorian calendar, established in 1582. But this calendar was not adopted by England (and the American colonies) until 1752, when Wednesday, September 2, was followed by Thursday, September 14. Let's verify that September 14, 1752, was indeed a Thursday. Since 2052 has a year code of 2 (from page 218, or by 52 + 13 − 63 = 2), then 1752 has a year code of 0. Thus September 14, 1752, has

Month Code + Date + Year Code = 4 + 14 + 0 = 18

and 18 − 14 = 4, so it was indeed a Thursday. However, our formula will not work for earlier dates (which were governed by the Julian calendar).

Finally, we remark that under the Gregorian calendar, a leap year occurs every four years, with the exception of years that are divisible by 100, although there is an exception to the exception: years divisible by 400 are leap years. Thus 1600, 2000, 2400, and 2800 are leap years, but 1700, 1800, 1900, 2100, 2200, 2300, and 2500 are not leap years. In fact, the Gregorian calendar repeats every 400 years, so you can convert any future date into a date near 2000. For example, March 19, 2361, and March 19, 2761, will have the same day of the week as March 19, 1961, which, as we already determined, is a Sunday.

EXERCISE: A DAY FOR ANY DATE

Determine the days of the week for the following dates:

1. January 19, 2007
2. February 14, 2012
3. June 20, 1993
4. September 1, 1983
5. September 8, 1954
6. November 19, 1863
7. July 4, 1776
8. February 22, 2222
9. June 31, 2468
10. January 1, 2358

Chapter ∞

Epilogue: How Math Helps Us Think About Weird Things

by Michael Shermer

As the publisher of *Skeptic* magazine, the executive director of the Skeptics Society, and a *Scientific American* editor with a monthly column entitled "Skeptic," I receive volumes of mail from people who challenge me with stories about their unusual experiences, such as haunted houses, ghosts, near-death and out-of-body experiences, UFO sightings, alien abductions, death-premonition dreams, and much more.

The most interesting stories to me are those about highly improbable events. The implication of the letter writer's tale is that if I cannot offer a satisfactory *natural* explanation for *that particular event,* the general principle of *supernaturalism* is preserved. A common story is the one about having a dream about the death of a friend or relative, then a phone call comes the next day about the unexpected death of the person in the dream. What are the odds of that? I am asked.

Here is where math comes in to play in our thinking and reasoning. I don't want to pontificate about how mathematics in school teaches students to think critically, because that has probably been said by nearly every math teacher in nearly every math class in nearly every school in America, at least once a year. I want to give some specific examples of how I use very simple math to help me on the job in explaining why weird things happen to people.

Although I cannot always explain such specific occurrences, a principle of probability called the Law of Large Numbers shows that an event with a low probability of occurrence in a small number of trials has a high probability of occurrence in a large number of trials. Or, as I like to say, million-to-one odds happen 295 times a day in America.

Let's begin with death premonitions. Here is a little "back-of-the-envelope" calculation I did. Psychologists tell us that the average person has about five dreams per day, which equals 1,825 dreams per year. Even if we remember only one out of ten dreams, that still results in 182.5 remembered dreams a year. There are 295 million Americans, so that means there will be 53.8 billion remembered dreams per year. Now, anthropologists and sociologists tell us that each of us knows about 150 people fairly well (that is, the average person has about 150 names in his or her address book about which can be said something significant). That means there is a network grid of 44.3 billion personal relationships among those 295 million Americans. The annual U.S. death rate from all causes across all ages is .008, or 2.6 million per year. It is inevitable that some of those 53.8 billion remembered dreams will be about some of these 2.6 million deaths among the 295 million Americans and their 44.3 billion relationships. *It would be a miracle, in fact, if some "death premonition" dreams did not come true.*

Even if my numbers are off, even way off, the point still stands. What are the odds of a death premonition dream coming true? Pretty darn good.

There is an additional psychological factor at work here called the *confirmation bias,* where we notice the hits and ignore the misses in support of our favorite beliefs. The confirmation bias explains how conspiracy theories work, for example. People who adhere to a particular conspiracy theory (9/11 was orchestrated by the Bush administration in order to launch a war in the Middle East), will look for and find little factoids here and there that seem to indicate that it might be true (Bush sat in that classroom reading to the children about goats as if he knew he was safe), while ignoring the vast body of evidence that points to another more likely explanation (Osama bin Laden and his band of international terrorists orchestrated 9/11). The confirmation bias also helps explain how astrologers, tarot-card readers, and psychics seem so successful at "reading" people. People who get a reading are likely to remember the handful of hits and forget the countless misses. When such hits and misses are actually counted—which I once did for an ABC television special on psychics—it turns out that there is nothing more than guessing and random chance at work.

In the case of the death-premonition dream, if just a couple of these people who have such dreams recount their miraculous tales in a public forum (next on *Oprah*!), the paranormal seems vindicated. In fact, it is nothing more than the laws of probability writ large.

This mathematical process of thinking about weird things led me to another back-of-the-envelope calculation about miracles. People typically invoke the term *miracle* to describe really unusual events, events whose odds of occurring are a "million to one." Okay, let's take that as our benchmark definition. A miracle is an event whose odds of occurrence are a million to one. Now, as we go about our day, we see and hear things happen about once per second. That is, data from the world and events around us are pouring in through our senses at a rate of about one per second. If we are awake and alert and out in the world for, say, eight hours a day, that means there are thirty thousand bits of data per day, or one million events per month that we take in. The vast majority of these data and events are completely meaningless, of course, and our brains are wired to filter out and forget the vast majority of them because we would be overwhelmed otherwise. But, in the course of a month, we would expect million-to-one odds to happen at least once. Add to that the confirmation bias where we will remember the most unusual events and forget all the rest, and it is inevitable that someone somewhere will report a miracle every month. And the tabloids will be there to record it!

This is a short primer on how science works. In our quest to understand how the world works, we need to determine what is real and what is not, what happens by chance and what happens because of some particular predictable cause. The problem we face is that the human brain was designed by evolution to pay attention to the really unusual events and ignore the vast body of data flowing by; as such, thinking statistically and with probabilities does not come naturally. Science, to that extent, does not come naturally. It takes some training and practice.

In addition, there are those pesky cognitive biases I mentioned, such as the confirmation bias. And there are others. The data do not just speak for themselves. Data are filtered through very subjective and biased brains. The *self-serving bias*, for example, dictates that we tend to see ourselves in a more positive light than others see us: national surveys show that most business people believe they are more moral than other business people, while psychologists who study moral intuition think they are more moral than other such psychologists. In one College Entrance Examination Board survey of 829,000 high school seniors, 0 percent rated themselves below average in "ability to get along with others," while 60 percent put themselves in the top 10 percent (presumably not all were from Lake Woebegone). And according to a 1997 *U.S. News & World Report* study on who Americans believe are most likely to go to heaven, 52 percent said Bill Clinton, 60 percent thought Princess Diana, 65 percent chose Michael Jordan, 79 percent selected Mother Teresa, and, at 87 percent, the person most likely to go to heaven was the survey taker!

Princeton University psychology professor Emily Pronin and her colleagues tested a bias called *blind spot,* in which subjects recognized the existence and influence in others of eight different cognitive biases, but they failed to see those same biases in themselves. In one study on Stanford University students, when asked to compare themselves to their peers on such personal qualities as friendliness and selfishness, they predictably rated themselves higher. Even when the subjects were warned about the *better-than-average bias* and were asked to reevaluate their original assessments, 63 percent claimed that their initial evaluations were objective, and 13 percent even claimed that they were originally too modest! In a second study, Pronin randomly assigned subjects high or low scores on a "social intelligence" test. Unsurprisingly, those given the high marks rated the test fairer and more useful than those receiving low marks. When asked if it was possible that they had been influenced by the score on the test, subjects responded that *other* participants had been far more biased than they were. In a third study in which Pronin queried subjects about what method they used to assess their own and others' biases, she found that people tend to use general theories of behavior when evaluating others, but use introspection when appraising themselves; however, in what is called the *introspection illusion,* people do not believe that others can be trusted to do the same. Okay for me but not for thee.

The University of California at Berkeley psychologist Frank J. Sulloway and I made a similar discovery of an *attribution bias* in a study we conducted on why people say they believe in God, and why they think other people believe in God. In general, most people attribute their own belief in God to such intellectual reasons as the good design and complexity of the world, whereas they attribute others' belief in God to such emotional reasons as it is comforting, gives meaning, and that they were raised to believe. Political scientists have made a similar discovery about political attitudes, where Republicans justify their conservative attitudes with rational arguments but claim that Democrats are "bleeding-heart liberals," and where Democrats claim that their liberal attitudes are the most rational but claim that Republicans are "heartless."

How does science deal with such subjective biases? How do we know when a claim is bogus or real? We want to be open-minded enough to accept radical new ideas when they occasionally come along, but we don't want to be so open-minded that our brains fall out. This problem led us at the Skeptics Society to create an educational tool called the Baloney Detection Kit, inspired by Carl Sagan's discussion of how to detect "baloney" in his marvelous book *The Demon-Haunted World*. In this Baloney Detection Kit, we suggest ten questions to ask when encountering any claim that can help us decide if we are being too open-minded in accepting it or too closed-minded in rejecting it.

1. How reliable is the source of the claim? As Daniel Kevles showed so effectively in his 1999 book *The Baltimore Affair,* in investigating possible scientific fraud there is a boundary problem in detecting a fraudulent signal within the background noise of mistakes and sloppiness that is a normal part of the scientific process. The investigation of research notes in a laboratory affiliated with Nobel laureate David Baltimore by an independent committee established by Congress to investigate potential fraud revealed a surprising number of mistakes. But science is messier than most people realize. Baltimore was exonerated when it became clear that there was no purposeful data manipulation.

2. Does this source often make similar claims? Pseudoscientists have a habit of going well beyond the facts, so when individuals make numerous extraordinary claims, they may be more than just iconoclasts. This is a matter of quantitative scaling, since some great thinkers often go beyond the data in their creative speculations. Cornell's Thomas Gold is notorious for his radical ideas, but he has been right often enough that other scientists listen to what he has to say. Gold proposes, for example, that oil is not a fossil fuel at all, but the by-product of a deep hot biosphere. Hardly any earth scientists I have spoken with take this thesis seriously, yet they do not consider Gold a crank. What we are looking for here is a pattern of fringe thinking that consistently ignores or distorts data.

3. Have the claims been verified by another source? Typically pseudoscientists will make statements that are unverified, or verified by a source within their own belief circle. We must ask who is checking the claims, and even who is checking the checkers. The biggest problem with the cold fusion debacle, for example, was not that scientists Stanley Pons and Martin Fleischman were wrong; it was that they announced their spectacular discovery before it was verified by other laboratories (at a press

conference no less), and, worse, when cold fusion was not repli-
cated, they continued to cling to their claim.

**4. How does the claim fit with what we know about how the
world works?** An extraordinary claim must be placed into a
larger context to see how it fits. When people claim that the
pyramids and the Sphinx were built more than ten thousand
years ago by an advanced race of humans, they are not present-
ing any context for that earlier civilization. Where are the rest of
the artifacts of those people? Where are their works of art, their
weapons, their clothing, their tools, their trash? This is simply
not how archaeology works.

**5. Has anyone gone out of the way to disprove the claim, or has
only confirmatory evidence been sought?** This is the confirmation bias, or the tendency to seek confirmatory evidence and
reject or ignore disconfirmatory evidence. The confirmation bias
is powerful and pervasive and is almost impossible for any of us
to avoid. It is why the methods of science that emphasize checking and rechecking, verification and replication, and especially
attempts to falsify a claim are so critical.

**6. Does the preponderance of evidence converge to the
claimant's conclusion, or a different one?** The theory of evolution, for example, is proven through a convergence of evidence
from a number of independent lines of inquiry. No one fossil, no
one piece of biological or paleontological evidence has "evolution" written on it; instead there is a convergence of evidence
from tens of thousands of evidentiary bits that adds up to a
story of the evolution of life. Creationists conveniently ignore
this convergence, focusing instead on trivial anomalies or cur-
rently unexplained phenomena in the history of life.

7. Is the claimant employing the accepted rules of reason and tools of research, or have these been abandoned in favor of others that lead to the desired conclusion? UFOlogists suffer this fallacy in their continued focus on a handful of unexplained atmospheric anomalies and visual misperceptions by eyewitnesses, while conveniently ignoring the fact that the vast majority (90 to 95 percent) of UFO sightings are fully explicable with prosaic answers.

8. Has the claimant provided a different explanation for the observed phenomena, or is it strictly a process of denying the existing explanation? This is a classic debate strategy—criticize your opponent and never affirm what you believe in order to avoid criticism. But this stratagem is unacceptable in science. Big Bang skeptics, for example, ignore the convergence of evidence of this cosmological model, focus on the few flaws in the accepted model, and have yet to offer a viable cosmological alternative that carries a preponderance of evidence in favor of it.

9. If the claimant has proffered a new explanation, does it account for as many phenomena as the old explanation? The HIV-AIDS skeptics argue that lifestyle, not HIV, causes AIDS. Yet, to make this argument they must ignore the convergence of evidence in support of HIV as the causal vector in AIDS, and simultaneously ignore such blatant evidence as the significant correlation between the rise in AIDS among hemophiliacs shortly after HIV was inadvertently introduced into the blood supply. On top of this, their alternative theory does not explain nearly as much of the data as the HIV theory.

10. Do the claimants' personal beliefs and biases drive the conclusions, or vice versa? All scientists hold social, political, and ideological beliefs that could potentially slant their interpretations of the data, but how do those biases and beliefs affect their research? At some point, usually during the peer-review system, such biases and beliefs are rooted out, or the paper or book is rejected for publication. This is why one should not work in an intellectual vacuum. If you don't catch the biases in your research, someone else will.

There is no definitive set of criteria we can apply in determining how open-minded we should be when encountering new claims and ideas, but with mathematical calculations on the odds of weird things happening and with an analysis of the sorts of questions we should ask when we encounter weird things, we have made a start toward coming to grips with our weird and wonderful world.

Answers

CHAPTER 1: A LITTLE GIVE AND TAKE

Two-Digit Addition (page 15)

 1. $23 + 16 = 23 + 10 + 6 = 33 + 6 = 39$

 2. $64 + 43 = 64 + 40 + 3 = 104 + 3 = 107$

 3. $95 + 32 = 95 + 30 + 2 = 125 + 2 = 127$

 4. $34 + 26 = 34 + 20 + 6 = 54 + 6 = 60$

 5. $89 + 78 = 89 + 70 + 8 = 159 + 8 = 167$

 6. $73 + 58 = 73 + 50 + 8 = 123 + 8 = 131$

 7. $47 + 36 = 47 + 30 + 6 = 77 + 6 = 83$

 8. $19 + 17 = 19 + 10 + 7 = 29 + 7 = 36$

 9. $55 + 49 = 55 + 40 + 9 = 95 + 9 = 104$

10. $39 + 38 = 39 + 30 + 8 = 69 + 8 = 77$

Three-Digit Addition (page 20)

 1. $242 + 137 = 242 + 100 + 30 + 7 = 342 + 30 + 7 = 372 + 7 = 379$

 2. $312 + 256 = 312 + 200 + 50 + 6 = 512 + 50 + 6 = 562 + 6 = 568$

 3. $635 + 814 = 635 + 800 + 10 + 4 = 1435 + 10 + 4 = 1445 + 4 = 1449$

 4. $457 + 241 = 457 + 200 + 40 + 1 = 657 + 40 + 1 = 697 + 1 = 698$

 5. $912 + 475 = 912 + 400 + 70 + 5 = 1312 + 70 + 5 = 1382 + 5 = 1387$

6. $852 + 378 = 852 + 300 + 70 + 8 = 1152 + 70 + 8 = 1222 + 8 = 1230$

7. $457 + 269 = 457 + 200 + 60 + 9 = 657 + 60 + 9 = 717 + 9 = 726$

8. $878 + 797 = 878 + 700 + 90 + 7 = 1578 + 90 + 7 = 1668 + 7 = 1675$ or $878 + 797 = 878 + 800 - 3 = 1678 - 3 = 1675$

9. $276 + 689 = 276 + 600 + 80 + 9 = 876 + 80 + 9 = 956 + 9 = 965$

10. $877 + 539 = 877 + 500 + 30 + 9 = 1377 + 30 + 9 = 1407 + 9 = 1416$

11. $5400 + 252 = 5400 + 200 + 52 = 5600 + 52 = 5652$

12. $1800 + 855 = 1800 + 800 + 55 = 2600 + 55 = 2655$

13. $6120 + 136 = 6120 + 100 + 30 + 6 = 6220 + 30 + 6 = 6250 + 6 = 6256$

14. $7830 + 348 = 7830 + 300 + 40 + 8 = 8130 + 40 + 8 = 8170 + 8 = 8178$

15. $4240 + 371 = 4240 + 300 + 70 + 1 = 4540 + 70 + 1 = 4610 + 1 = 4611$

Two-Digit Subtraction (page 23)

1. $38 - 23 = 38 - 20 - 3 = 18 - 3 = 15$

2. $84 - 59 = 84 - 60 + 1 = 24 + 1 = 25$

3. $92 - 34 = 92 - 40 + 6 = 52 + 6 = 58$

4. $67 - 48 = 67 - 50 + 2 = 17 + 2 = 19$

5. $79 - 29 = 79 - 20 - 9 = 59 - 9 = 50$ or $79 - 29 = 79 - 30 + 1 = 49 + 1 = 50$

6. $63 - 46 = 63 - 50 + 4 = 13 + 4 = 17$

7. $51 - 27 = 51 - 30 + 3 = 21 + 3 = 24$

8. $89 - 48 = 89 - 40 - 8 = 49 - 8 = 41$

9. $125 - 79 = 125 - 80 + 1 = 45 + 1 = 46$

10. $148 - 86 = 148 - 90 + 4 = 58 + 4 = 62$

Three-Digit Subtraction (page 28)

1. $583 - 271 = 583 - 200 - 70 - 1 = 383 - 70 - 1 = 313 - 1 = 312$

2. $936 - 725 = 936 - 700 - 20 - 5 = 236 - 20 - 5 = 216 - 5 = 211$

3. $587 - 298 = 587 - 300 + 2 = 287 + 2 = 289$

4. $763 - 486 = 763 - 500 + 14 = 263 + 14 = 277$

5. $204 - 185 = 204 - 200 + 15 = 04 + 15 = 19$

6. $793 - 402 = 793 - 400 - 2 = 393 - 2 = 391$

7. $219 - 176 = 219 - 200 + 24 = 19 + 24 = 43$

8. $978 - 784 = 978 - 800 + 16 = 178 + 16 = 194$

9. $455 - 319 = 455 - 400 + 81 = 55 + 81 = 136$

10. $772 - 596 = 772 - 600 + 4 = 172 + 4 = 176$

11. $873 - 357 = 873 - 400 + 43 = 473 + 43 = 516$

12. $564 - 228 = 564 - 300 + 72 = 264 + 72 = 336$

13. $1428 - 571 = 1428 - 600 + 29 = 828 + 29 = 857$

14. $2345 - 678 = 2345 - 700 + 22 = 1645 + 22 = 1667$

15. $1776 - 987 = 1776 - 1000 + 13 = 776 + 13 = 789$

CHAPTER 2: PRODUCTS OF A MISSPENT YOUTH

2-by-1 Multiplication (page 35)

1.	2.	3.	4.	5.
82	43	67	71	93
× 9	× 7	× 5	× 3	× 8
720	280	300	210	720
+ 18	+ 21	+ 35	+ 3	+ 24
738	301	335	213	744

6.		7.	8.	9.	10.
49 *or*	49	28	53	84	58
× 9	× 9	× 4	× 5	× 5	× 6
360	450	80	250	400	300
+ 81	− 9	+ 32	+ 15	+ 20	+ 48
441	441	112	265	420	348

11.	97	12.	78	13.	96	14.	75	15.	57
	× 4		× 2		× 9		× 4		× 7
	360		140		810		280		350
	+ 28		+ 6		+ 54		+ 20		+ 49
	388		156		864		300		399

16.	37	17.	46	18.	76	19.	29	20.	64
	× 6		× 2		× 8		× 3		× 8
	180		80		560		60		480
	+ 42		+ 12		+ 48		+ 27		+ 32
	222		92		608		87		512

3-by-1 Multiplication (page 43)

1.	431	2.	637	3.	862	4.	957
	× 6		× 5		× 4		× 6
	2400		3000		3200		5400
	+ 180		+ 150		+ 240		+ 300
	2580		3150		3440		5700
	+ 6		+ 35		+ 8		+ 42
	2586		3185*		3448		5742

5.	927	6.	728	7.	328	8.	529
	× 7		× 2		× 6		× 9
	6300		1400		1800		4500
	+ 140		+ 40		+ 120		+ 180
	6440		1440		1920		4680
	+ 49		+ 16		+ 48		+ 81
	6489		1456		1968		4761

9.	807	10.	587	11.	184	12.	214
	× 9		× 4		× 7		× 8
	7200		2000		700		1600
	+ 63		+ 320		+ 560		+ 80
	7263		2320		1260		1680
			+ 28		+ 28		+ 32
			2348*		1288		1712

*With this kind of problem you can easily say the answer out loud as you go.

13.
```
   757
 ×   8
  5600
 + 400
  6000
 +  56
  6056
```

14.
```
   259
 ×   7
  1400
 + 350
  1750
 +  63
  1813
```

15.
```
   297
 ×   8
  1600
 + 720
  2320
 +  56
  2376
```

```
          297
  or    ×   8
300 × 8 = 2400
− 3 × 8 = −  24
         2376
```

16.
```
   751
 ×   9
  6300
 + 450
  6750
 +   9
  6759
```

17.
```
   457
 ×   7
  2800
 + 350
  3150
 +  49
  3199
```

18.
```
   339
 ×   8
  2400
 + 240
  2640
 +  72
  2712
```

19.
```
   134
 ×   8
   800
 + 240
  1040
 +  32
  1072
```

20.
```
   611
 ×   3
  1800
 +  33
  1833
```

21.
```
   578
 ×   9
  4500
 + 630
  5130
 +  72
  5202
```

22.
```
   247
 ×   5
  1000
 + 200
  1200
 +  35
  1235*
```

23.
```
   188
 ×   6
   600
 + 480
  1080
 +  48
  1128
```

24.
```
   968
 ×   6
  5400
 + 360
  5760
 +  48
  5808
```

25.
```
   499
 ×   9
  3600
 + 810
  4410
 +  81
  4491
```

```
          499
  or    ×   9
500 × 9 = 4500
− 1 × 9 = −   9
         4491
```

26.
```
   670
 ×   4
  2400
 + 280
  2680
```

27.
```
   429
 ×   3
  1200
 +  60
  1260
 +  27
  1287
```

28.
```
   862
 ×   5
  4000
 + 300
  4300
 +  10
  4310*
```

*With this kind of problem you can easily say the answer out loud as you go.

29.	285	30.	488	31.	693	32.	722
	× 6		× 9		× 6		× 9
	1200		3600		3600		6300
	+ 480		+ 720		+ 540		+ 180
	1680		4320		4140		6480
	+ 30		+ 72		+ 18		+ 18
	1710		4392		4158		6498

33.	457	34.	767	35.	312	36.	691
	× 9		× 3		× 9		× 3
	3600		2100		2700		1800
	+ 450		+ 180		+ 90		+ 270
	4050		2280		2790		2070
	+ 63		+ 21		+ 18		+ 3
	4113		2301		2808		2073

Two-Digit Squares (page 48)

1. 14^2
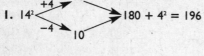
$180 + 4^2 = 196$

2. 27^2
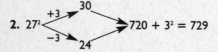
$720 + 3^2 = 729$

3. 65^2

$4200 + 5^2 = 4225$

4. 89^2

$7920 + 1^2 = 7921$

5. 98^2 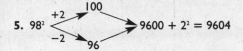 $9600 + 2^2 = 9604$

6. 31^2 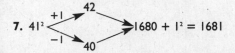 $960 + 1^2 = 961$

7. 41^2 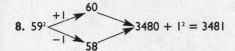 $1680 + 1^2 = 1681$

8. 59^2 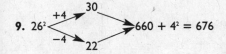 $3480 + 1^2 = 3481$

9. 26^2  $660 + 4^2 = 676$

10. 53^2 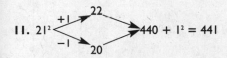 $2800 + 3^2 = 2809$

11. 21^2  $440 + 1^2 = 441$

12. 64^2 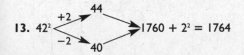 $4080 + 4^2 = 4096$

13. 42^2 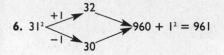 $1760 + 2^2 = 1764$

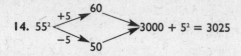

14. 55^2 +5 → 60, −5 → 50 → $3000 + 5^2 = 3025$

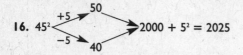

15. 75^2 +5 → 80, −5 → 70 → $5600 + 5^2 = 5625$

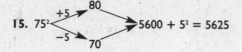

16. 45^2 +5 → 50, −5 → 40 → $2000 + 5^2 = 2025$

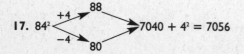

17. 84^2 +4 → 88, −4 → 80 → $7040 + 4^2 = 7056$

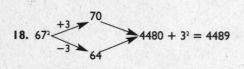

18. 67^2 +3 → 70, −3 → 64 → $4480 + 3^2 = 4489$

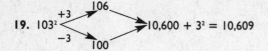

19. 103^2 +3 → 106, −3 → 100 → $10,600 + 3^2 = 10,609$

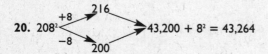

20. 208^2 +8 → 216, −8 → 200 → $43,200 + 8^2 = 43,264$

CHAPTER 3: NEW AND IMPROVED PRODUCTS

Multiplying by 11 (page 58)

1. 35   3___5   = 385
    × 11   8

2. 48   4___8   = 528
    × 11   12

3. 94   9___4   = 1034
    × 11   13

2-by-2 Addition-Method Multiplication Problems (page 59)

1.

$$
\begin{array}{r}
31\ (30+1) \\
\times\ 41 \\
\end{array}
$$

$$
\begin{array}{rr}
30 \times 41 = & 1230 \\
1 \times 41 = & +\ \ 41 \\
\hline
& 1271
\end{array}
$$

or

$$
\begin{array}{r}
31 \\
\times\ \ 41\ (40+1) \\
\end{array}
$$

$$
\begin{array}{rr}
40 \times 31 = & 1240 \\
1 \times 31 = & +\ \ 31 \\
\hline
& 1271
\end{array}
$$

2.

$$
\begin{array}{r}
27\ (20+7) \\
\times\ 18 \\
\end{array}
$$

$$
\begin{array}{rr}
20 \times 18 = & 360 \\
7 \times 18 = & +\ 126 \\
\hline
& 486
\end{array}
$$

3.

$$
\begin{array}{r}
59\ (50+9) \\
\times\ 26 \\
\end{array}
$$

$$
\begin{array}{rr}
50 \times 26 = & 1300 \\
9 \times 26 = & +\ 234 \\
\hline
& 1534
\end{array}
$$

4.

$$
\begin{array}{r}
53\ (50+3) \\
\times\ 58 \\
\end{array}
$$

$$
\begin{array}{rr}
50 \times 58 = & 2900 \\
3 \times 58 = & +\ 174 \\
\hline
& 3074
\end{array}
$$

5.

$$
\begin{array}{r}
77 \\
\times\ \ 43\ (40+3) \\
\end{array}
$$

$$
\begin{array}{rr}
40 \times 77 = & 3080 \\
3 \times 77 = & +\ 231 \\
\hline
& 3311
\end{array}
$$

6.

$$
\begin{array}{r}
23\ (20+3) \\
\times\ 84 \\
\end{array}
$$

$$
\begin{array}{rr}
20 \times 84 = & 1680 \\
3 \times 84 = & +\ 252 \\
\hline
& 1932
\end{array}
$$

or

$$
\begin{array}{r}
23 \\
\times\ \ 84\ (80+4) \\
\end{array}
$$

$$
\begin{array}{rr}
80 \times 23 = & 1840 \\
4 \times 23 = & +\ \ 92 \\
\hline
& 1932
\end{array}
$$

7.

$$
\begin{array}{r}
62\ (60+2) \\
\times\ 94 \\
\end{array}
$$

$$
\begin{array}{rr}
60 \times 94 = & 5640 \\
2 \times 94 = & +\ 188 \\
\hline
& 5828
\end{array}
$$

8.

$$
\begin{array}{r}
88\ (80+8) \\
\times\ 76 \\
\end{array}
$$

$$
\begin{array}{rr}
80 \times 76 = & 6080 \\
8 \times 76 = & +\ 608 \\
\hline
& 6688
\end{array}
$$

9.

$$
\begin{array}{r}
92\ (90+2) \\
\times\ 35 \\
\end{array}
$$

$$
\begin{array}{rr}
90 \times 35 = & 1230 \\
2 \times 35 = & +\ \ 70 \\
\hline
& 3220
\end{array}
$$

10.

$$
\begin{array}{r}
34 \\
\times\ 11 \\
\hline
\end{array}
\qquad 3\_\_\_\_4 = 374\ or\quad
\begin{array}{r}
340 \\
+\ 34 \\
\hline
374
\end{array}
$$

$$\ \ \ \ \ 7$$

11.

$$
\begin{array}{r}
85 \\
\times\ 11 \\
\hline
\end{array}
\qquad 8\_\_\_\_5 = 935\ or\quad
\begin{array}{r}
850 \\
+\ 85 \\
\hline
935
\end{array}
$$

$$\ \ \ \ \ 13$$

2-by-2 Subtraction-Method Multiplication Problems (page 63)

1.
$$
\begin{array}{r}
29 \ (30 - 1) \\
\times \quad 45 \\
\hline
30 \times 45 = \quad 1350 \\
-1 \times 45 = - \quad 45 \\
\hline
1305
\end{array}
$$

2.
$$
\begin{array}{r}
98 \ (100 - 2) \\
\times \quad 43 \\
\hline
100 \times 43 = \quad 4300 \\
-2 \times 43 = - \quad 86 \\
\hline
4214
\end{array}
$$

3.
$$
\begin{array}{r}
47 \\
\times \quad 59 \ (60 - 1) \\
\hline
60 \times 47 = \quad 2820 \\
-1 \times 47 = - \quad 47 \\
\hline
2773
\end{array}
$$

4.
$$
\begin{array}{r}
68 \ (70 - 2) \\
\times \quad 43 \\
\hline
70 \times 38 = \quad 2660 \\
-2 \times 38 = - \quad 76 \\
\hline
2584
\end{array}
$$

5.
$$
\begin{array}{r}
96 \ (100 - 4) \\
\times \quad 29 \\
\hline
100 \times 29 = \quad 2900 \\
-4 \times 29 = - \quad 116 \\
\hline
2784
\end{array}
$$

or
$$
\begin{array}{r}
96 \\
\times \quad 29 \ (30 - 1) \\
\hline
30 \times 96 = \quad 2880 \\
-1 \times 96 = - \quad 96 \\
\hline
2784
\end{array}
$$

6.
$$
\begin{array}{r}
79 \ (80 - 1) \\
\times \quad 54 \\
\hline
80 \times 54 = \quad 4320 \\
-1 \times 54 = - \quad 54 \\
\hline
4266
\end{array}
$$

7.
$$
\begin{array}{r}
37 \\
\times \quad 19 \ (20 - 1) \\
\hline
20 \times 37 = \quad 740 \\
-1 \times 37 = - \quad 37 \\
\hline
703
\end{array}
$$

8.
$$
\begin{array}{r}
87 \ (90 - 3) \\
\times \quad 22 \\
\hline
90 \times 22 = \quad 1980 \\
-3 \times 22 = - \quad 66 \\
\hline
1914
\end{array}
$$

9.
$$
\begin{array}{r}
85 \\
\times \quad 38 \ (40 - 2) \\
\hline
40 \times 85 = \quad 3400 \\
-2 \times 85 = - \quad 170 \\
\hline
3230
\end{array}
$$

10.
$$
\begin{array}{r}
57 \\
\times \quad 39 \ (40 - 1) \\
\hline
40 \times 57 = \quad 2280 \\
-1 \times 57 = - \quad 57 \\
\hline
2223
\end{array}
$$

11.
$$
\begin{array}{r}
88 \\
\times \quad 49 \ (50 - 1) \\
\hline
50 \times 88 = \quad 4400 \\
-1 \times 88 = - \quad 88 \\
\hline
4312
\end{array}
$$

2-by-2 Factoring-Method Multiplication Problems (page 68)

1. $27 \times 14 = 27 \times 7 \times 2 = 189 \times 2 = 378$ *or* $14 \times 27 = 14 \times 9 \times 3 = 126 \times 3 = 378$

2. $86 \times 28 = 86 \times 7 \times 4 = 602 \times 4 = 2408$

3. $57 \times 14 = 57 \times 7 \times 2 = 399 \times 2 = 798$

4. $81 \times 48 = 81 \times 8 \times 6 = 648 \times 6 = 3888$ *or* $48 \times 81 = 48 \times 9 \times 9 = 432 \times 9 = 3888$

5. $56 \times 29 = 29 \times 7 \times 8 = 203 \times 8 = 1624$

6. $83 \times 18 = 83 \times 6 \times 3 = 498 \times 3 = 1494$

7. $72 \times 17 = 17 \times 9 \times 8 = 153 \times 8 = 1224$

8. $85 \times 42 = 85 \times 6 \times 7 = 510 \times 7 = 3570$

9. $33 \times 16 = 33 \times 8 \times 2 = 264 \times 2 = 528$ *or* $16 \times 33 = 16 \times 11 \times 3 = 176 \times 3 = 528$

10. $62 \times 77 = 62 \times 11 \times 7 = 682 \times 7 = 4774$

11. $45 \times 36 = 45 \times 6 \times 6 = 270 \times 6 = 1620$ *or* $45 \times 36 = 45 \times 9 \times 4 = 405 \times 4 = 1620$ *or* $36 \times 45 = 36 \times 9 \times 5 = 324 \times 5 = 1620$ *or* $36 \times 45 = 36 \times 5 \times 9 = 180 \times 9 = 1620$

12. $37 \times 48 = 37 \times 8 \times 6 = 296 \times 6 = 1776$

2-by-2 General Multiplication—Anything Goes! (page 70)

1.
$$
\begin{array}{r}
53 \\
\times \ 39 \ (40 - 1) \\
\hline
40 \times 53 = \quad 2120 \\
-1 \times 53 = - \ 53 \\
\hline
2067
\end{array}
$$

or
$$
\begin{array}{r}
53 \ (50 + 3) \\
\times \ 39 \\
\hline
50 \times 39 = \quad 1950 \\
3 \times 39 = + \ 117 \\
\hline
2067
\end{array}
$$

2.
$$
\begin{array}{r}
81 \ (80 + 1) \\
\times \ 57 \\
\hline
80 \times 57 = \quad 4560 \\
1 \times 57 = + \ 57 \\
\hline
4617
\end{array}
$$

or $57 \times 81 = 57 \times 9 \times 9 = 513 \times 9 = 4617$

3.
$$
\begin{array}{r}
73 \\
\times \ 18 \ (9 \times 2)
\end{array}
$$

$73 \times 18 = 73 \times 9 \times 2 = 657 \times 2 = 1314$ *or* $73 \times 18 = 73 \times 6 \times 3 = 438 \times 3 = 1314$

4.

$$89 \,(90 - 1)$$
$$\underline{\times \ 55}$$
$$90 \times 55 = \quad 4950$$
$$1 \times 55 = \underline{- \ 55}$$
$$4895$$

or $89 \times 55 = 89 \times 11 \times 5 =$
$979 \times 5 = 4895$

5.

$$77$$
$$\underline{\times 36} \,(4 \times 9)$$

$77 \times 36 = 77 \times 4 \times 9 = 308 \times 9 = 2772$ or
$77 \times 36 = 77 \times 9 \times 4 = 693 \times 4 = 2772$

6.

$$92$$
$$\underline{\times \ 53} \,(50 + 3)$$
$$50 \times 92 = \quad 4600$$
$$3 \times 92 = \underline{+ \ 276}$$
$$4876$$

7. 87^2

$$87^2 \begin{array}{c} \overset{+3}{\longrightarrow} 90 \\ \overset{-3}{\longrightarrow} 84 \end{array} \longrightarrow 7560 + 3^2 = 7569$$

8.

$$67$$
$$\underline{\times \ 58} \,(60 - 2)$$
$$60 \times 67 = \quad 4020$$
$$-2 \times 67 = \underline{- \ 154}$$
$$3886$$

9.

$$56 \,(8 \times 7)$$
$$\underline{\times 37}$$

$37 \times 56 = 37 \times 8 \times 7 = 296 \times 7 = 2072$ or
$37 \times 56 = 37 \times 7 \times 8 = 259 \times 8 = 2072$

10.

$$59$$
$$\underline{\times \ 21} \,(20 + 1)$$
$$20 \times 59 = \quad 1180$$
$$1 \times 59 = \underline{+ \quad 59}$$
$$1239$$

or

$$59 \,(60 - 1)$$
$$\underline{\times \ 21}$$
$$60 \times 21 = \quad 1260$$
$$-1 \times 21 = \underline{- \quad 21}$$
$$1239$$

or $59 \times 21 = 59 \times 7 \times 3 = 413 \times 3 = 1239$

11.

$$37$$
$$\underline{\times 72} \,(9 \times 8)$$

$37 \times 9 \times 8 = 333 \times 8 = 2664$

12.

$$57$$
$$\underline{\times \ 73} \,(70 + 3)$$
$$70 \times 57 = \quad 3990$$
$$3 \times 57 = \underline{+ \ 171}$$
$$4161$$

13.   38
    $\underline{\times\ 63}$ (9×7)

$38 \times 63 = 38 \times 9 \times 7 =$
$342 \times 7 = 2394$

14.   43 $(40 + 3)$
                    $\underline{\times\ \ 76}$
$40 \times 76 =\quad 3040$
$3 \times 76 = \underline{+\ \ 228}$
                    3268

15.   43
    $\underline{\times\ 75}$ $(5 \times 5 \times 3)$

$43 \times 75 = 43 \times 5 \times 5 \times 3 =$
$215 \times 5 \times 3 = 1075 \times 3 = 3225$

16.   74
                    $\underline{\times\ \ 62}$ $(60 + 2)$
$60 \times 74 =\quad 4440$
$2 \times 74 = \underline{+\ 148}$
                    4588

17.   61 $(60 + 1)$
                    $\underline{\times\ \ 37}$
$60 \times 37 =\quad 2220$
$1 \times 37 = \underline{+\ \ 37}$
                    2257

18.   36 (6×6)
    $\underline{\times\ 41}$

$41 \times 36 = 41 \times 6 \times 6 =$
$246 \times 6 = 1476$

19.   54 (9×6)
    $\underline{\times\ 53}$

$54 \times 53 = 53 \times 9 \times 6 =$
$477 \times 6 = 2862$

20. 53^2 $\begin{array}{c} {\scriptstyle +3} \nearrow\ 56 \searrow \\ {\scriptstyle -3} \searrow\ 50 \nearrow \end{array}$ $2800 + 3^2 = 2809$

21.   83 $(80 + 3)$
                    $\underline{\times\ 58}$
$80 \times 58 =\quad 4640$
$3 \times 58 = \underline{+\ 174}$
                    4814

22.   91 $(90 + 1)$
                    $\underline{\times\ \ 46}$
$90 \times 46 =\quad 4140$
$1 \times 46 = \underline{+\ \ 46}$
                    4186

23.   52 $(50 + 2)$
                    $\underline{\times\ \ 47}$
$50 \times 47 =\quad 2350$
$2 \times 47 = \underline{+\ \ 94}$
                    2444

24.   29 $(30 - 1)$
                    $\underline{\times\ \ 26}$
$30 \times 26 =\quad 780$
$-1 \times 26 = \underline{-\ \ 26}$
                    754

25. 41
  × 15 (5 × 3)

  41 × 15 = 41 × 5 × 3 =
  205 × 3 = 615

26.   65
        × 19 (20 − 1)
  20 × 65 =   1300
  −1 × 65 = − 65
              1235

27. 34
  × 27 (9 × 3)

  34 × 27 = 34 × 9 × 3 =
  306 × 3 = 918

28.   69 (70 − 1)
        × 78
  70 × 78 =   5460
  −1 × 78 = − 78
              5382

29. 95
  × 81 (9 × 9)

  95 × 81 = 95 × 9 × 9 =
  855 × 9 = 7695

30.   65 (60 + 5)
        × 47
  60 × 47 =   2820
  5 × 47 = + 235
              3055

31.   65
        × 69 (70 − 1)
  70 × 65 =   4550
  −1 × 65 = − 65
              4485

32.   95
        × 26 (20 + 6)
  20 × 95 =   1900
  6 × 95 = + 570
              2470

33.   41 (40 + 1)
        × 93
  40 × 93 =   3720
  1 × 93 = + 93
              3813

Three-Digit Squares (page 76)

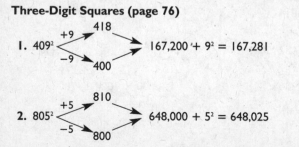

1. 409^2 +9 → 418
             −9 → 400   → $167,200 + 9^2 = 167,281$

2. 805^2 +5 → 810
             −5 → 800   → $648,000 + 5^2 = 648,025$

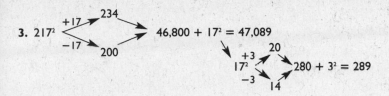

3. 217^2

+17 → 234
−17 → 200

$46,800 + 17^2 = 47,089$

17^2
+3 → 20
−3 → 14

$280 + 3^2 = 289$

4. 896^2

+4 → 900
−4 → 892

$802,800 + 4^2 = 802,816$

5. 345^2

+45 → 390
−45 → 300

$117,000 + 45^2 = 119,025$

45^2
+5 → 50
−5 → 40

$2,000 + 5^2 = 2025$

6. 346^2

+46 → 392
−46 → 300

$117,600 + 46^2 = 119,716$

46^2
+4 → 50
−4 → 42

$2,100 + 4^2 = 2116$

7. 276^2

+24 → 300
−24 → 252

$75,600 + 24^2 = 76,176$

24^2
+4 → 28
−4 → 20

$560 + 4^2 = 576$

8. 682^2

+18 → 700
−18 → 664

$464,800 + 18^2 = 465,124$

18^2
+2 → 20
−2 → 16

$320 + 2^2 = 324$

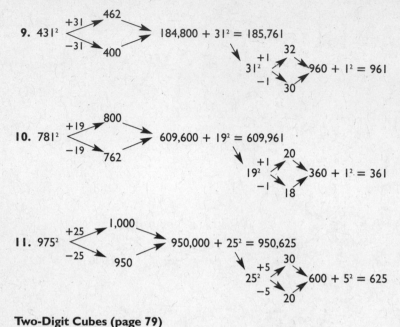

9. 431^2
+31 → 462
−31 → 400
$184,800 + 31^2 = 185,761$
31^2
+1 → 32
−1 → 30
$960 + 1^2 = 961$

10. 781^2
+19 → 800
−19 → 762
$609,600 + 19^2 = 609,961$
19^2
+1 → 20
−1 → 18
$360 + 1^2 = 361$

11. 975^2
+25 → 1,000
−25 → 950
$950,000 + 25^2 = 950,625$
25^2
+5 → 30
−5 → 20
$600 + 5^2 = 625$

Two-Digit Cubes (page 79)

1. $12^3 = (10 \times 12 \times 14) + (2^2 \times 12) = 1680 + 48 = 1728$
2. $17^3 = (14 \times 17 \times 20) + (3^2 \times 17) = 4760 + 153 = 4913$
3. $21^3 = (20 \times 21 \times 22) + (1^2 \times 21) = 9240 + 21 = 9261$
4. $28^3 = (26 \times 28 \times 30) + (2^2 \times 28) = 21,840 + 112 = 21,952$
5. $33^3 = (30 \times 33 \times 36) + (3^2 \times 33) = 35,640 + 297 = 35,937$
6. $39^3 = (38 \times 39 \times 40) + (1^2 \times 39) = 59,280 + 39 = 59,319$
7. $40^3 = 40 \times 40 \times 40 = 64,000$
8. $44^3 = (40 \times 44 \times 48) + (4^2 \times 44) = 84,480 + 704 = 85,184$
9. $52^3 = (50 \times 52 \times 54) + (2^2 \times 52) = 140,400 + 208 = 140,608$
10. $56^3 = (52 \times 56 \times 60) + (4^2 \times 56) = 174,720 + 896 = 175,616$
11. $65^3 = (60 \times 65 \times 70) + (5^2 \times 65) = 273,000 + 1,625 = 274,625$
12. $71^3 = (70 \times 71 \times 72) + (1^2 \times 71) = 357,840 + 71 = 357,911$
13. $78^3 = (76 \times 78 \times 80) + (2^2 \times 78) = 474,240 + 312 = 474,552$
14. $85^3 = (80 \times 85 \times 90) + (5^2 \times 85) = 612,000 + 2,125 = 614,125$
15. $87^3 = (84 \times 87 \times 90) + (3^2 \times 87) = 657,720 + 783 = 658,503$
16. $99^3 = (98 \times 99 \times 100) + (1^2 \times 99) = 970,200 + 99 = 970,299$

CHAPTER 4: DIVIDE AND CONQUER

One-Digit Division (page 84)

1. $35\frac{3}{9}$
    9)318
    $- 27$
    48
    $- 45$
    3

2. $145\frac{1}{5}$
    5)726
    $- 5$
    22
    $- 20$
    26
    $- 25$
    1

3. $61\frac{1}{7}$
    7)428
    $- 42$
    08
    $- 7$
    1

4. $36\frac{1}{8}$
    8)289
    $- 24$
    49
    $- 48$
    1

5. $442\frac{2}{3}$
    3)1328
    $- 12$
    $- 12$
    08
    $- 6$
    2

6. $695\frac{2}{4}$
    4)2782
    $- 24$
    38
    $- 36$
    22
    $- 20$
    2

Two-Digit Division (page 93)

1. $43\frac{7}{17}$
    17)738
    $- 68$
    58
    $- 51$
    7

2. $24\frac{15}{24}$
    24)591
    $- 48$
    111
    $- 96$
    15

3. $4\frac{5}{79}$
    79)321
    $- 316$
    5

4. $152\frac{12}{28}$
    28)4268
    $- 28$
    146
    $- 140$
    68
    $- 56$
    12

5. $655\frac{9}{11}$
    11)7214
    $- 66$
    61
    $- 55$
    64
    $- 55$
    9

6. $170\frac{14}{18}$
    18)3074
    $- 18$
    127
    $- 126$
    14

Decimalization (page 98)

1. $\frac{2}{5} = .40$ 2. $\frac{4}{7} = .571428$ 3. $\frac{3}{8} = .375$ 4. $\frac{9}{12} = .75$

5. $\frac{5}{12} = .4166$ 6. $\frac{6}{11} = .5454$ 7. $\frac{14}{24} = .5833$ 8. $\frac{13}{27} = .481$

9. $\frac{18}{48} = .375$ 10. $\frac{10}{14} = .714285$ 11. $\frac{6}{32} = .1875$ 12. $\frac{19}{45} = .422$

Testing for Divisibility (page 101)

Divisibility by 2

1. 53,42<u>8</u>	2. 29<u>3</u>	3. 724<u>1</u>	4. 984<u>6</u>
Yes	No	No	Yes

Divisibility by 4

5. 39<u>32</u>	6. 67,3<u>48</u>	7. 3<u>58</u>	8. 57,9<u>29</u>
Yes	Yes	No	No

Divisibility by 8

9. 59,3<u>66</u>	10. 73,<u>488</u>	11. <u>248</u>	12. 611<u>1</u>
No	Yes	Yes	No

Divisibility by 3

13. 83,671
No: 8 + 3 + 6 + 7 + 1 = 25

14. 94,737
Yes: 9 + 4 + 7 + 3 + 7 = 30

15. 7359
Yes: 7 + 3 + 5 + 9 = 24

16. 3,267,486
Yes: 3 + 2 + 6 + 7 + 4 + 8 + 6 = 36

Divisibility by 6

17. 533<u>4</u>
Yes: 5 + 3 + 3 + 4 = 15

18. 67,38<u>6</u>
Yes: 6 + 7 + 3 + 8 + 6 = 30

19. 248
No: 2 + 4 + 8 = 14

20. 599<u>1</u>
No: odd

Divisibility by 9

21. 1234
No: 1 + 2 + 3 + 4 = 10

22. 8469
Yes: 8 + 4 + 6 + 9 = 27

23. 4,425,575

No: 4 + 4 + 2 + 5 + 5 + 7 + 5 = 32

24. 314,159,265

Yes: 3 + 1 + 4 + 1 + 5 + 9 + 2 + 6 + 5 = 36

Divisibility by 5

25. 47,83<u>0</u>   **26.** 43,76<u>2</u>   **27.** 56,78<u>5</u>   **28.** 37,21<u>0</u>

Yes   No   Yes   Yes

Divisibility by 11

29. 53,867

Yes: 5 − 3 + 8 − 6 + 7 = 11

30. 4969

No: 4 − 9 + 6 − 9 = −8

31. 3828

Yes: 3 − 8 + 2 − 8 = −11

32. 941,369

Yes: 9 − 4 + 1 − 3 + 6 − 9 = 0

Divisibility by 7

33. 5784

No: 5784 − 7 = 5777

577 − 7 = 570

57

34. 7336

Yes: 7336 + 14 = 7350

735 − 35 = 700

7

35. 875

Yes: 875 − 35 = 840

84 − 14 = 70

7

36. 1183

Yes: 1183 + 7 = 1190

119 + 21 = 140

14

Divisibility by 17

37. 694

No: 694 − 34 = 660

66

38. 629

Yes: 629 + 51 = 680

68

39. 8273

No: 8273 + 17 = 8290

829 + 51 = 880

88

40. 13,855

Yes: 13,855 + 85 = 13,940

1394 − 34 = 1360

136 + 34 = 170

17

Multiplying Fractions (page 102)

1. $\dfrac{6}{35}$ 2. $\dfrac{44}{63}$ 3. $\dfrac{18}{28} = \dfrac{9}{14}$ 4. $\dfrac{63}{80}$

Dividing Fractions (page 103)

1. $\dfrac{4}{5}$ 2. $\dfrac{5}{18}$ 3. $\dfrac{10}{15} = \dfrac{2}{3}$

Simplifying Fractions (page 104)

1. $\dfrac{1}{3} = \dfrac{4}{12}$ 2. $\dfrac{5}{6} = \dfrac{10}{12}$ 3. $\dfrac{3}{4} = \dfrac{9}{12}$ 4. $\dfrac{5}{2} = \dfrac{30}{12}$

5. $\dfrac{8}{10} = \dfrac{4}{5}$ 6. $\dfrac{6}{15} = \dfrac{2}{5}$ 7. $\dfrac{24}{36} = \dfrac{2}{3}$ 8. $\dfrac{20}{36} = \dfrac{5}{9}$

Adding Fractions (Equal Denominators) (page 105)

1. $\dfrac{2}{9} + \dfrac{5}{9} = \dfrac{7}{9}$ 2. $\dfrac{5}{12} + \dfrac{4}{12} = \dfrac{9}{12} = \dfrac{3}{4}$

3. $\dfrac{5}{18} + \dfrac{6}{18} = \dfrac{11}{18}$ 4. $\dfrac{3}{10} + \dfrac{3}{10} = \dfrac{6}{10} = \dfrac{3}{5}$

Adding Fractions (Unequal Denominators) (page 106)

1. $\dfrac{1}{5} + \dfrac{1}{10} = \dfrac{2}{10} + \dfrac{1}{10} = \dfrac{3}{10}$ 2. $\dfrac{1}{6} + \dfrac{5}{18} = \dfrac{3}{18} + \dfrac{5}{18} = \dfrac{8}{18} = \dfrac{4}{9}$

3. $\dfrac{1}{3} + \dfrac{1}{5} = \dfrac{5}{15} + \dfrac{3}{15} = \dfrac{8}{15}$ 4. $\dfrac{2}{7} + \dfrac{5}{21} = \dfrac{6}{21} + \dfrac{5}{21} = \dfrac{11}{21}$

5. $\dfrac{2}{3} + \dfrac{3}{4} = \dfrac{8}{12} + \dfrac{9}{12} = \dfrac{17}{12}$ 6. $\dfrac{3}{7} + \dfrac{3}{5} = \dfrac{15}{35} + \dfrac{21}{35} = \dfrac{36}{35}$

7. $\dfrac{2}{11} + \dfrac{5}{9} = \dfrac{18}{99} + \dfrac{55}{99} = \dfrac{73}{99}$

Subtracting Fractions (page 107)

1. $\dfrac{8}{11} - \dfrac{3}{11} = \dfrac{5}{11}$ 2. $\dfrac{12}{7} - \dfrac{8}{7} = \dfrac{4}{7}$

3. $\dfrac{13}{18} - \dfrac{5}{18} = \dfrac{8}{18} = \dfrac{4}{9}$ 4. $\dfrac{4}{5} - \dfrac{1}{15} = \dfrac{12}{15} - \dfrac{1}{15} = \dfrac{11}{15}$

5. $\dfrac{9}{10} - \dfrac{3}{5} = \dfrac{9}{10} - \dfrac{6}{10} = \dfrac{3}{10}$ 6. $\dfrac{3}{4} - \dfrac{2}{3} = \dfrac{9}{12} - \dfrac{8}{12} = \dfrac{1}{12}$

7. $\dfrac{7}{8} - \dfrac{1}{16} = \dfrac{14}{16} - \dfrac{1}{16} = \dfrac{13}{16}$

8. $\dfrac{4}{7} - \dfrac{2}{5} = \dfrac{20}{35} - \dfrac{14}{35} = \dfrac{6}{35}$

9. $\dfrac{8}{9} - \dfrac{1}{2} = \dfrac{16}{18} - \dfrac{9}{18} = \dfrac{7}{18}$

CHAPTER 5: GOOD ENOUGH

Addition Guesstimation (page 128)

Exact

1.	1479	2.	57,293	3.	312,025	4.	8,971,011
	+ 1105		+ 37,421		+ 79,419		+ 4,016,367
	2584		94,714		391,444		12,987,378

Guesstimates

1. 1500 *or* 1480
    + 1100   + 1100
    2600   2580

2. 57,000 *or* 57,300
    + 37,000   + 37,400
    94,000   94,700

3. 310,000 *or* 312,000
    + 80,000   + 79,000
    390,000   391,000

4. 9 million *or* 8.9 million
    + 4 million   + 4.0 million
    13 million   12.9 million

    or 8.97 million
    + 4.02 million
    12.99 million

Exact	*Guesstimates*
$ 2.67	$ 2.50
1.95	2.00
7.35	7.50
9.21	9.00
0.49	0.50
11.21	11.00
0.12	0.00
6.14	6.00
8.31	8.50
$47.35	$47.00

Answers

Subtraction Guesstimation (page 129)
Exact

1.	4926	**2.**	67,221	**3.**	526,978	**4.**	8,349,241
	− 1659		− 9,874		− 42,009		− 6,103,839
	3267		57,347		484,969		2,245,402

Guesstimates

1. 4900
    − 1700
     3200

2. 67,000 *or* 67,200
     − 10,000   − 9,900
      57,000   57,300

3. 530,000 *or* 527,000
     − 40,000   − 42,000
      490,000   485,000

4. 8.3 million *or* 8.35 million
     − 6.1 million   − 6.10 million
      2.2 million   2.25 million

Division Guesstimation (page 129)
Exact

1. 625.57
   7)4379

2. 4,791.6
   5)23,958

3. 42,247.15
   13)549,213

4.   17,655.21
   289)5,102,357

5.   40.90
   203,637)8,329,483

Guesstimates

1. 630
   7)4400

2. 4,800
   5)24,000

3. 42,000
   13)550,000

4.   17,000
   ≈ 300)5,100,000 = 3)51,000

5.   40
   ≈ 200,000)8,000,000 = 200)8000

Multiplication Guesstimation (page 129)
Exact

1. 98
   × 27
   2646

2. 76
   × 42
   3192

3. 88
   × 88
   7744

4. 539
   × 17
   9163

5. 312
   × 98
   30,576

6. 639
   × 107
   68,373

7. 428
   × 313
   133,964

8. 51,276
   × 489
   25,073,964

9.   104,972
× 11,201
1,175,791,372

10.   5,462,741
× 203,413
1,111,192,535,033

Guesstimates

1. ·100
× 25
2500

2. 78
× 40
3120

3. 90
× 86
7740

4. 540
× 17
9180

5. 310
× 100
31,000

6. 646 *or* 640
× 100   × 110
64,600   70,400

7. 430
× 310
133,300

8. 51,000·
× 490
24,990,000

9. 105,000
× 11,000
1155 million
= 1.155 billion

10. 5,500,000
× 200,00
1100 billion
= 1.1 trillion

Square Root Guesstimation (page 130)
Exact (to two decimal places)

1. $\dfrac{4.12}{\sqrt{17}}$ **2.** $\dfrac{5.91}{\sqrt{35}}$ **3.** $\dfrac{12.76}{\sqrt{163}}$ **4.** $\dfrac{65.41}{\sqrt{4279}}$ **5.** $\dfrac{89.66}{\sqrt{8039}}$

Divide and Average

1. $4\overline{)17}\,^{4.2}$ $\dfrac{4+4.2}{2}=4.1$

2. $6\overline{)35}\,^{5.8}$ $\dfrac{6+5.8}{2}=5.9$

3. $10\overline{)163}\,^{16.3}$ $\dfrac{10+16.3}{2}=13.15$

4. $60\overline{)4279}\,^{71}$ $\dfrac{60+71}{2}=65.5$

5. $90\overline{)8039}\,^{89}$ $\dfrac{90+89}{2}=89.5$

Everyday Math (page 130)

1. $8.80 + $4.40 = $13.20

2. $5.30 + $2.65 = $7.95

3. $74 ÷ 2 ÷ 2 = $37 ÷ 2 = $18.50

4. Since 70 ÷ 10 = 7, seven years

5. Since 70 ÷ 6 = 11.67, it will take twelve years to double

6. Since 110 ÷ 7 = 15.714, it will take sixteen years to triple

7. Since 70 ÷ 7 = 10, it will take ten years to double, then another ten years to double again. Thus it will take twenty years to quadruple.

8. $M = \dfrac{\$100,000(0.0075)(1.0075)^{120}}{(1.00333)^{120} - 1} = \dfrac{\$750(2.451)}{1.451} = \$1267$

9. $M = \dfrac{\$30,000(0.004167)(1.004167)^{42}}{(1.004167)^{42} - 1} = \dfrac{\$125(1.22)}{0.22} = \$693$

CHAPTER 6: MATH FOR THE BOARD

Columns of Numbers (page 149)

1.			2.		
672	→	6	$ 21.56	→	5
1,367	→	8	19.38	→	3
107	→	8	211.02	→	6
7,845	→	6	9.16	→	7
358	→	7	26.17	→	7
210	→	3	1.43	→	3
+ 916	→	7	$288.72	→	9
11,475	→	9			

Subtracting on Paper (page 150)

1.			2.		
75,423	→	3	876,452	→	5
− 46,298	→	2	− 593,876	→	2
29,125	→	1	282,576	→	3

3.			4.		
3,249,202	→	4	45,394,358	→	5
− 2,903,445	→	9	− 36,472,659	→	6
345,757	→	4	8,921,699	→	8

Square Root Guesstimation (page 150)

1.
$$3.\ 8\ 7$$
$$\sqrt{15.0000}$$
$3^2 = \underline{9}$
600
$6\underline{8} \times \underline{8} = \underline{544}$
5600
$76\underline{7} \times \underline{7} = 5369$

2.
$$2\ 2.\ 4\ 0$$
$$\sqrt{502.0000}$$
$2^2 = \underline{4}$
102
$4\underline{2} \times \underline{2} = \underline{84}$
1800
$44\underline{4} \times \underline{4} = \underline{1776}$
2400
$448\underline{0} \times \underline{0} = \quad 0$

3.
$$2\ 0.\ 9\ 5$$
$$\sqrt{439.2000}$$
$2^2 = \underline{4}$
039
$4\underline{0} \times \underline{0} = \underline{0}$
3920
$40\underline{9} \times \underline{9} = \underline{3681}$
23900
$448\underline{0} \times \underline{0} = 20925$

4.
$$1\ 9 \text{ exactly}$$
$$\sqrt{361}$$
$1^2 = \underline{1}$
261
$2\underline{9} \times \underline{9} = \underline{261}$
0

Pencil-and-Paper Multiplication (page 150)

1.
54 → 9
× 37 → 1
1998 → 9

2.
273 → 3
× 217 → 1
59,241 → 3

3.
725 → 5
× 609 → 6
441,525 → 3

4.
3,309 → 6
× 2,868 → 6
9,490,212 → 9

5.
52,819 → 7
× 47,820 → 3
2,525,804,580 → 3

6.
3,923,759 → 3
× 2,674,093 → 4
10,492,496,475,587 → 3

CHAPTER 8: THE TOUGH STUFF MADE EASY

Four-Digit Squares (page 169)

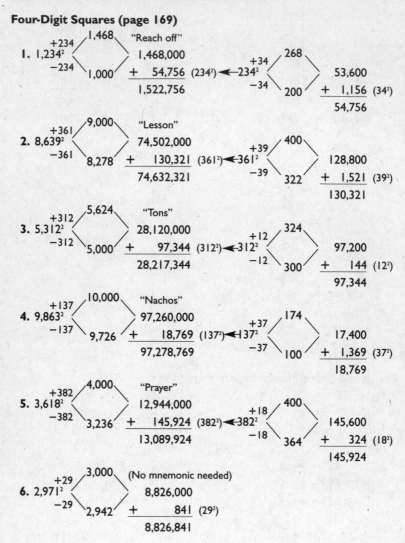

1. $1,234^2$

+234 ⟍ 1,468 ⟍ "Reach off"
−234 ⟋ 1,000 ⟋

1,468,000
+ 54,756 (234^2)
1,522,756

+34 ⟍ 268 ⟍
−34 ⟋ 200 ⟋ ← 234^2

53,600
+ 1,156 (34^2)
54,756

2. $8,639^2$

+361 ⟍ 9,000 ⟍ "Lesson"
−361 ⟋ 8,278 ⟋

74,502,000
+ 130,321 (361^2)
74,632,321

+39 ⟍ 400 ⟍
−39 ⟋ 322 ⟋ ← 361^2

128,800
+ 1,521 (39^2)
130,321

3. $5,312^2$

+312 ⟍ 5,624 ⟍ "Tons"
−312 ⟋ 5,000 ⟋

28,120,000
+ 97,344 (312^2)
28,217,344

+12 ⟍ 324 ⟍
−12 ⟋ 300 ⟋ ← 312^2

97,200
+ 144 (12^2)
97,344

4. $9,863^2$

+137 ⟍ 10,000 ⟍ "Nachos"
−137 ⟋ 9,726 ⟋

97,260,000
+ 18,769 (137^2)
97,278,769

+37 ⟍ 174 ⟍
−37 ⟋ 100 ⟋ ← 137^2

17,400
+ 1,369 (37^2)
18,769

5. $3,618^2$

+382 ⟍ 4,000 ⟍ "Prayer"
−382 ⟋ 3,236 ⟋

12,944,000
+ 145,924 (382^2)
13,089,924

+18 ⟍ 400 ⟍
−18 ⟋ 364 ⟋ ← 382^2

145,600
+ 324 (18^2)
145,924

6. $2,971^2$

+29 ⟍ 3,000 ⟍ (No mnemonic needed)
−29 ⟋ 2,942 ⟋

8,826,000
+ 841 (29^2)
8,826,841

3-by-2 Exercises Using Factoring, Addition, and Subtraction Methods (page 175)

1.
$$858$$
$$\times\ 15\ (5 \times 3)$$
$$858 \times 15 = 858 \times 5 \times 3 =$$
$$4,290 \times 3 = 12,870$$

2.
$$796\ (800 - 4)$$
$$\times\quad 19$$
$$800 \times 19 =\quad 15,200$$
$$-4 \times 19 = -\quad 76$$
$$15,124$$

3.
$$148$$
$$\times\quad 62\ (60 + 2)$$
$$148 \times 60 =\quad 8,880$$
$$148 \times 2 = +\quad 296$$
$$9,176$$

or
$$148\ (74 \times 2)$$
$$\times\ 62\ (60 + 2)$$
$$62 \times 148 = 62 \times 74 \times 2 =$$
$$4,588 \times 2 = 9,176$$

4.
$$773$$
$$\times\ 42\ (7 \times 6)$$
$$773 \times 42 = 773 \times 7 \times 6 =$$
$$5,411 \times 6 = 32,466$$

5.
$$906\ (900 + 6)$$
$$\times\quad 46$$
$$900 \times 46 =\quad 41,400$$
$$6 \times 46 = +\quad 276$$
$$41,676$$

6.
$$952\ (950 + 2)$$
$$\times\quad 26$$
$$950 \times 26 =\quad 24,700$$
$$2 \times 26 = +\quad 52$$
$$24,752$$

7.
$$411\ (410 + 1)$$
$$\times\quad 93$$
$$410 \times 93 =\quad 38,130$$
$$1 \times 93 = +\quad 93$$
$$38,223$$

8.
$$967$$
$$\times\quad 51\ (50 + 1)$$
$$50 \times 967 =\quad 48,350$$
$$1 \times 967 = +\quad 967$$
$$49,317$$

9.
$$484$$
$$\times\ 75\ (5 \times 5 \times 3)$$
$$484 \times 75 = 484 \times 5 \times 5 \times 3 =$$
$$2,420 \times 5 \times 3 = 12,100 \times 3 =$$
$$36,300$$

10.
$$126 \ (9 \times 7 \times 2)$$
$$\underline{\times \ 87}$$

$126 \times 87 = 87 \times 9 \times 7 \times 2 =$
$783 \times 7 \times 2 = 5{,}481 \times 2 =$
$10{,}962$

11.
$$157$$
$$\underline{\times \ 33} \ (11 \times 3)$$

$157 \times 33 = 157 \times 11 \times 3 =$
$1727 \times 3 = 5181$

12.
$$616 \ (610 + 6)$$
$$\underline{\times \quad 37}$$
$610 \times 37 = \quad 22{,}570$
$6 \times 37 = \underline{+ \quad 222}$
$$22{,}792$$

13.
$$841$$
$$\underline{\times \ 72} \ (9 \times 8)$$

$841 \times 72 = 841 \times 9 \times 8 =$
$7{,}569 \times 8 = 60{,}552$

14.
$$361 \ (360 + 1)$$
$$\underline{\times \quad 41}$$
$360 \times 41 = \quad 14{,}760$
$1 \times 41 = \underline{+ \quad 41}$
$$14{,}801$$

15.
$$218$$
$$\underline{\times \quad 68} \ (70 - 2)$$
$70 \times 218 = \quad 15{,}260$
$-2 \times 218 = \underline{- \quad 436}$
$$14{,}824$$

16.
$$538 \ (540 - 2) \quad or$$
$$\underline{\times \quad 53}$$
$540 \times 53 = \quad 28{,}620$
$-2 \times 53 = \underline{- \quad 106}$
$$28{,}514$$

$$538 \ (530 + 8)$$
$$\underline{\times \quad 53}$$
$530 \times 53 = \quad 28{,}090$
$8 \times 53 = \underline{+ \quad 424}$
$$28{,}514$$

17.
$$817$$
$$\underline{\times \quad 61} \ (60 + 1)$$
$60 \times 817 = \quad 49{,}020$
$1 \times 817 = \underline{+ \quad 817}$
$$49{,}837$$

18.
$$668$$
$$\underline{\times \ 63} \ (9 \times 7)$$

$668 \times 63 = 668 \times 9 \times 7 =$
$6{,}012 \times 7 = 42{,}084$

19.
$$499 \ (500 - 1)$$
$$\underline{\times \quad 25}$$
$500 \times 25 = \quad 12{,}500$
$-1 \times 25 = \underline{- \quad 25}$
$$12{,}475$$

20.
$$144$$
$$\underline{\times \ 56} \ (7 \times 8)$$

$144 \times 56 = 144 \times 7 \times 8 =$
$1008 \times 8 = 8064$

21.
$$281$$
$$\times\ 44\ (11 \times 4)$$
or
$$281\ (280 + 1)$$
$$\times\ \ \ \ 44$$

$281 \times 44 = 281 \times 11 \times 4 =$
$280 \times 44 = \ \ \ 12,320$

$3,091 \times 4 = 12,364$
$1 \times 44 = \underline{+\ \ \ \ \ 44}$
$$12,364$$

22.
$$988\ (1000 - 12)$$
$$\times\ \ \ \ 22$$

$1000 \times 22 = \ \ \ 22,000$

$-12 \times 22 = \underline{-\ \ \ \ 264}$

$$21,736$$

23.
$$383$$
$$\times\ 49\ (7 \times 7)$$

$383 \times 49 = 383 \times 7 \times 7 =$

$2,681 \times 7 = 18,767$

24.
$$589\ (600 - 11)$$
$$\times\ \ \ \ 87$$

$600 \times 87 = \ \ \ 52,200$

$11 \times -87 = \underline{-\ \ \ 957}$

$$51,243$$

25.
$$286$$
$$\times\ 64\ (8 \times 8)$$

$286 \times 64 = 286 \times 8 \times 8 =$

$2,288 \times 8 = 18,304$

26.
$$853$$
$$\underline{\times\ 32}\ (8 \times 4)$$

$853 \times 32 = 853 \times 8 \times 4 =$

$6,824 \times 4 = 27,296$

27.
$$878$$
$$\underline{\times\ 24}\ (8 \times 3)$$

$878 \times 24 = 878 \times 8 \times 3 =$

$7,024 \times 3 = 21,072$

28.
$$423\ (47 \times 9)$$
$$\underline{\times\ 65}$$

$423 \times 65 = 65 \times 47 \times 9 =$

$3,055 \times 9 = 27,495$

29.
$$154\ (11 \times 14)$$
$$\underline{\times\ 19}$$

$154 \times 19 = 19 \times 11 \times 14 =$

$209 \times 7 \times 2 = 1463 \times 2 =$

2926

30.
$$834\ (800 + 34)$$
$$\times\ \ \ \ 34$$

$800 \times 34 = \ \ \ 27,200$

$34 \times 34 = \underline{+\ 1,156}$

$$28,356$$

31.
$$545$$
$$\underline{\times\ 27}\ (9 \times 3)$$

$545 \times 27 = 545 \times 9 \times 3 =$

$4,905 \times 3 = 14,715$

32.

$$653 \ (650 + 3)$$
$$\underline{\times \quad 69}$$

$650 \times 69 = \quad 44{,}850$
$3 \times 69 = \underline{+ \quad 207}$
$\qquad\qquad\quad 45{,}057$

33.

$$216 \ (6 \times 6 \times 6)$$
$$\underline{\times \ 78}$$

$216 \times 78 = 78 \times 6 \times 6 \times 6 =$
$468 \times 6 \times 6 = 2{,}808 \times 6 =$
$16{,}848$

34.

$$822$$
$$\underline{\times \quad 95} \ (100 - 5)$$

$100 \times 822 = \quad 82{,}200$
$-5 \times 822 = \underline{- \ 4{,}110}$
$\qquad\qquad\quad 78{,}090$

Five-Digit Squares (page 180)

1. $45{,}795^2$

$$795 \ (800 - 5)$$
$$\underline{\times \quad 45}$$

$800 \times 45 = \quad 36{,}000$
$-5 \times 45 = \underline{- \quad 225}$ 　　　　"Lilies"
$\qquad\qquad 35{,}775 \times 2{,}000 = 71{,}550{,}000$

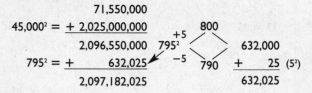

$\qquad\qquad\qquad\quad 71{,}550{,}000$
$45{,}000^2 = \underline{+ \ 2{,}025{,}000{,}000}$
$\qquad\qquad\quad 2{,}096{,}550{,}000$
$795^2 = \underline{+ \qquad 632{,}025}$
$\qquad\qquad\quad 2{,}097{,}182{,}025$

2. $21{,}231^2$

$$231$$
$$\underline{\times \quad 21} \ (7 \times 3)$$

$231 \times 7 \times 3 = 1{,}617 \times 3 = 4{,}851$

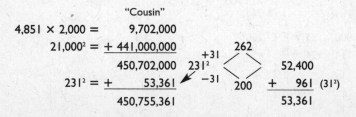

　　　　　　　"Cousin"
$4{,}851 \times 2{,}000 = \qquad 9{,}702{,}000$
$21{,}000^2 = \underline{+ \ 441{,}000{,}000}$
$\qquad\qquad\quad 450{,}702{,}000$
$231^2 = \underline{+ \qquad 53{,}361}$
$\qquad\qquad\quad 450{,}755{,}361$

3. $58,324^2$

$$324 \ (9 \times 6 \times 6)$$
$$\times \quad 58$$

$324 \times 58 = 58 \times 9 \times 6 \times 6 = 522 \times 6 \times 6 =$
$3,132 \times 6 = 18,792$

"Liver"

$18,792 \times 2,000 =$	$37,584,000$
$58,000^2 = +$	$3,364,000,000$
	$3,401,584,000$
$324^2 = +$	$104,976$
	$3,401,688,976$

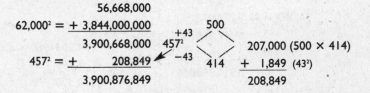

$+24$ 348
324^2 $104,400 \ (348 \times 300)$
-24 300 $+ \quad 576 \ (24^2)$
        $104,976$

4. $62,457^2$

$$457$$
$$\times \quad 62 \ (60 + 2)$$
$60 \times 457 = \quad 27,420$
$2 \times 457 = + \quad 914$ "Judge off"
$28,334 \times 2,000 = 56,668,000$

	$56,668,000$
$62,000^2 = +$	$3,844,000,000$
	$3,900,668,000$
$457^2 = +$	$208,849$
	$3,900,876,849$

$+43$ 500
457^2 $207,000 \ (500 \times 414)$
-43 414 $+ \ 1,849 \ (43^2)$
        $208,849$

5. $89,854^2$

$$854$$
$$\times \quad 89 \ (90 - 1)$$
$90 \times 854 = \quad 76,860$
$-1 \times 854 = - \quad 854$ "Stone"
$$76,006 \times 2,000 = 152,012,000$$

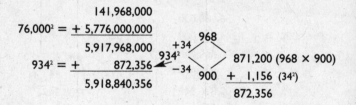

$$152,012,000$$
$89,000^2 = \ + \ 7,921,012,000$
$$8,073,012,000 \quad 854^2$$
$854^2 = \ + \qquad 729,316$
$$8,073,741,316$$

900
+46
808
−46
854^2
727,200 (900 × 808)
+ 2,116 (46^2)
729,316

6. $76,934^2$

$$934 \ (930 + 4)$$
$$\times \quad 76$$
$930 \times 76 = \quad 70,680$
$4 \times 76 = + \quad 304$ "Pie Chief"
$$76,984 \times 2,000 = 141,968,000$$

$$141,968,000$$
$76,000^2 = \ + \ 5,776,000,000$
$$5,917,968,000 \quad 934^2$$
$934^2 = \ + \qquad 872,356$
$$5,918,840,356$$

968
+34
900
−34
871,200 (968 × 900)
+ 1,156 (34^2)
872,356

3-by-3 Multiplication (page 192)

1.

$$644 \; (640 + 4) \quad or \quad 644 \; (7 \times 92)$$
$$\times \quad 286 \qquad\qquad\qquad \times \; 286$$

$640 \times 286 = \quad 183{,}040 \; (8 \times 8 \times 10)$

$4 \times 200 = \underline{+ \quad\quad 800}$ $\qquad 684 \times 286 = 286 \times 7 \times 92 =$

$\qquad\qquad\quad 183{,}840$ $\qquad\quad 2{,}002 \times 92 = 184{,}184$

$4 \times 86 = \underline{+ \quad\quad 344}$

$\qquad\qquad\quad 184{,}184$

2.

$$596 \; (600 - 4)$$
$$\times \quad 167$$

$600 \times 167 = \quad 100{,}200$

$-4 \times 167 = \underline{- \quad\quad 668}$

$\qquad\qquad\quad 99{,}532$

3.

$$853$$
$$\times \quad 325 \; (320 + 5)$$

$320 \times 853 = \quad 272{,}960$

$5 \times 800 = \underline{+ \; 4{,}000}$

$\qquad\qquad\quad 276{,}960$

$5 \times 53 = \underline{+ \quad\; 265}$

$\qquad\qquad\quad 277{,}225$

4.

$$343 \; (7 \times 7 \times 7)$$
$$\times \quad 226$$

$343 \times 226 = 226 \times 7 \times 7 \times 7 =$

$1{,}582 \times 7 \times 7 = 11{,}074 \times 7 =$

$77{,}518$

5.

$$809 \; (800 + 9)$$
$$\times \quad 527$$

$800 \times 527 = \quad 421{,}600$

$9 \times 527 = \underline{+ \quad 4{,}743}$

$\qquad\qquad\quad 426{,}343$

6.

$$942 \; (+42)$$
$$\times \quad 879 \; (-21)$$

$900 \times 921 = \quad 828{,}900$

$-21 \times 42 = \underline{- \quad\; 882}$

$\qquad\qquad\quad 828{,}018$

7.

$$692 \ (+8)$$
$$\times \quad 644 \ (-56)$$

$$700 \times 636 = \quad 445,200$$
$$(-8) \times (-56) = + \quad \underline{448}$$
$$445,648$$

8.

$$446$$
$$\times \quad 176 \ (11 \times 8 \times 2)$$

$446 \times 176 = 446 \times 11 \times 8 \times 2 =$
$4,906 \times 8 \times 2 = 39,248 \times 2 =$
$78,496$

9.

$$658 \ (47 \times 7 \times 2)$$
$$\times \quad 468 \ (52 \times 9)$$

$658 \times 468 = 52 \times 47 \times 9 \times 7 \times 2 =$
$2,444 \times 9 \times 7 \times 2 = 21,996 \times 7 \times 2 =$
$153,972 \times 2 = 307,944$

10.

$$273 \ (91 \times 3)$$
$$\times \quad 138 \ (46 \times 3)$$

$273 \times 138 = 91 \times 46 \times 9 =$
$4,186 \times 9 = 37,674$

11.

$$824$$
$$\times \quad 206 \ (412^2)$$

$$400 \times 424 = \quad 169,600$$
$$12 \times 12 = + \quad \underline{144}$$
$$169,744$$

12.

$$642 \ (107 \times 6)$$
$$\times 249 \ (83 \times 3)$$

$642 \times 249 = 107 \times 83 \times 18 = 8,881 \times 9 \times 2 =$
$79,929 \times 2 = 159,858$

13.   783 (87 × 9)
     × 589

   783 × 589 = 589 × 87 × 9 = 51,243 × 9 = 461,187

14.   871 (−29)
                 ×   926 (+26)
     900 × 897 =   807,300
     −29 × 26 = −   754
                    806,546

15.   341
              ×   715
     7 × 341 =   2,387
     3 × 15 = +   45
                 2,432 × 100 =   243,200
                    41 × 15 = +   615
                                243,815

16.   417
                 ×   298 (300 − 2)
     300 × 417 =   125,100
     −2 × 417 = −   834
                    124,266

17.   557
     × 756 (9 × 84)

   557 × 756 = 557 × 9 × 84 = 5,013 × 7 × 6 × 2 =
   35,091 × 6 × 2 = 210,546 × 2 = 421,092

18.   976 (1000 − 24)
                 ×   878
     878 × 1,000 =   878,000
     −878 × 24 = −   21,072
                    856,928

19.   765
    × 350 (7 × 5 × 10)

    765 × 350 = 765 × 7 × 5 × 10 = 5,355 × 5 × 10 =
    26,775 × 10 = 267,750

20.   154 (11 × 14)
    × 423 (47 × 9)

    154 × 423 = 47 × 11 × 9 × 14 = 517 × 9 × 7 × 2 =
    4,653 × 2 × 7 = 9,306 × 7 = 65,142

21.   545 (109 × 5)
                 ×   834
    100 × 834 =   83,400
      9 × 834 = + 7,506
                 90,906 × 5 = 454,530

22.   216 (6 × 6 × 6)
    × 653

    216 × 653 = 653 × 6 × 6 × 6 =
    3,918 × 6 × 6 = 23,508 × 6 = 141,048

23.   393 (400 − 7)
                 ×   822
    400 × 822 =   328,800
    −7 × 822 = −   5,754
                 323,046

5-by-5 Multiplication (page 198)

1. 65,154
    × 19,423

                    "Neck ripple"

    423 × 65 =   27,495
    154 × 19 = + 2,926   "Mouse round"

                30,421 × 1,000 =   30,421,000
            65 × 19 × 1 million = + 1,235,000,000
                                        1,265,421,000
                    154 × 423 = +   65,142
                                        1,265,486,142

2. 34,545
    × 27,834

                    "Knife mulch"

    834 × 34 =   28,356
    545 × 27 = +14,715   "Room scout"

                43,071 × 1,000 =   43,071,000
            34 × 27 × 1 million = +   918,000,000
                                        961,071,000
                    834 × 545 = +   454,530
                                        961,525,530

3. 69,216
    × 78,653

                    "Roll silk"

    653 × 69 =   45,057
    216 × 78 = +16,848   "Shoot busily"

                61,905 × 1,000 =   61,905,000
            69 × 78 × 1 million = + 5,382,000,000
                                        5,443,905,000
                    216 × 653 = +   141,048
                                        5,444,046,048

4.   95,393
    $\times$ 81,822

"Cave soups"

822 $\times$ 95 =   78,090
393 $\times$ 81 = +31,833   "Toss-up Panama"
            109,923 $\times$ 1,000 =   109,923,000
    95 $\times$ 81 $\times$ 1 million = + 7,804,000,000
                        7,804,923,000
            393 $\times$ 822 = +   323,046
                        7,805,246,046

A Day for Any Date (page 221)

 1. January 19, 2007, is Friday: 6 + 19 + 1 = 26; 26 − 21 = 5

 2. February 14, 2012, is Tuesday: 1 + 14 + 1 = 16; 16 − 14 = 2

 3. June 20, 1993, is Sunday: 3 + 5 + 20 = 28; 28 − 28 = 0

 4. September 1, 1983, is Thursday: 4 + 1 + 6 = 11; 11 − 7 = 4

 5. September 8, 1954, is Wednesday: 4 + 8 + 5 = 17; 17 − 14 = 3

 6. November 19, 1863, is Thursday: 2 + 19 + 4 = 25; 25 − 21 = 4

 7. July 4, 1776, is Thursday: 5 + 4 + 2 = 11; 11 − 7 = 4

 8. February 22, 2222, is Friday: 2 + 22 + 2 = 26; 26 − 21 = 5

 9. June 31, 2468, doesn't exist (only 30 days in June)! But June 30, 2468, is Saturday, so the next day would be Sunday.

10. January 1, 2358, is Wednesday: 6 + 1 + 3 = 10; 10 − 7 = 3

Bibliography

RAPID CALCULATION

Cutler, Ann, and Rudolph McShane. *The Trachtenberg Speed System of Basic Mathematics.* New York: Doubleday, 1960.

Devi, Shakuntala. *Figuring: The Joys of Numbers.* New York: Basic Books, 1964.

Doerfler, Ronald W. *Dead Reckoning: Calculating Without Instruments.* Houston: Gulf Publishing Company, 1993.

Flansburg, Scott, and Victoria Hay. *Math Magic.* New York: William Morrow and Co., 1993.

Handley, Bill. *Speed Mathematics: Secrets of Lightning Mental Calculation.* Queensland, Australia: Wrightbooks, 2003.

Julius, Edward H. *Rapid Math Tricks and Tips: 30 Days to Number Power.* New York: John Wiley & Sons, 1992.

Lucas, Jerry. *Becoming a Mental Math Wizard.* Crozet, Virginia: Shoe Tree Press, 1991.

Menninger, K. *Calculator's Cunning.* New York: Basic Books, 1964.

Smith, Steven B. *The Great Mental Calculators: The Psychology, Methods, and Lives of Calculating Prodigies, Past and Present.* New York: Columbia University Press, 1983.

Sticker, Henry. *How to Calculate Quickly.* New York: Dover, 1955.

Stoddard, Edward. *Speed Mathematics Simplified.* New York: Dover, 1994.

Tirtha, Jagadguru Swami Bharati Krishna, Shankaracharya of Govardhana Pitha. *Vedic Mathematics or "Sixteen Simple Mathematical Formulae from the Vedas."* Banaras, India: Hindu University Press, 1965.

MEMORY

Lorayne, Harry, and Jerry Lucas. *The Memory Book*. New York: Ballan-
tine Books, 1974.

Sanstrom, Robert. *The Ultimate Memory Book*. Los Angeles: Stepping
Stone Books, 1990.

RECREATIONAL MATHEMATICS

Gardner, Martin. *Magic and Mystery*. New York: Random House, 1956.

———. *Mathematical Carnival*. Washington, D.C.: Mathematical Association of America, 1965.

———. *Mathematical Magic Show*. New York: Random House, 1977.

———. *The Unexpected Hanging and Other Mathematical Diversions*.
New York: Simon & Schuster, 1969.

Huff, Darrell. *How to Lie with Statistics*. New York: Norton, 1954.

Paulos, John Allen. *Innumeracy: Mathematical Illiteracy and Its Conse-
quences*. New York: Hill and Wang, 1988.

Stewart, Ian. *Game, Set, and Math: Enigmas and Conundrums*. New York:
Penguin Books, 1989.

ADVANCED MATHEMATICS (BY ARTHUR BENJAMIN)

Benjamin, Arthur T., and Jennifer J. Quinn. *Proofs That Really Count: The
Art of Combinatorial Proof*. Washington: Mathematical Association of
America, 2003.

Benjamin, Arthur T., and Kan Yasuda. "Magic 'Squares' Indeed!," *The
American Mathematical Monthly* 106, no. 2 (February 1999): 152–56.

Index

addition:
  an "amazing" sum (Randi),
    212–14
  carrying a number in, 13–14
  columns of numbers (on paper),
    132–33; exercises, 149;
    answers, 256
  fractions with equal
    denominators, 104–5;
    exercises, 105; answers, 252
  fractions with unequal
    denominators, 105–6;
    exercises, 106; answers, 252
  guesstimation, 108–11; at the
    supermarket, 111; exercises,
    128; answers, 253
  leapfrog, 203–6
  left to right, 6–8, 11–21
  in multiplying three-by-three
    numbers, 188–90; exercises,
    192–93; answers,
    265–68
  in multiplying three-by-two
    numbers, 172–73; exercises,
    175–76; answers, 259–62
  in multiplying two-digit
    numbers, 54–59; exercises, 59;
    answers, 241
  with overlapping digits, 19
  three-digit numbers, 15–21;
    exercises, 19–21; answers,
    233–34

three-digit to four-digit numbers,
    18–19
  two-digit numbers, 12–15;
    exercises, 15; answers, 233
Aitken, Alexander Craig, 153
"amazing" sum (Randi), 212–14
associative law of multiplication,
    65
astrology, 224
attribution bias, 227

Baloney Detection Kit, 227–31
Baltimore, David, 228
Baltimore Affair, The (Kevles),
    227–28
belief in God, 227
better-than-average bias, 226
Bidder, George Parker, 109
Big Bang theory, 230
blind spots, 226

casting out elevens, 145, 147–49
casting out nines, 133, 134
chance, random, 224, 225
checking solutions:
  by casting out elevens, 145,
    147–49
  by casting out nines, 133, 134
  digital roots method, 133
  mod sums method, 133–34,
    135
Chevalier, Auguste, 121

About the Authors

DR. ARTHUR BENJAMIN is a professor of mathematics at Harvey Mudd College in Claremont, California, having received his PhD in mathematical sciences from Johns Hopkins University in 1989. In 2000, the Mathematical Association of America awarded him the Haimo Prize for Distinguished College Teaching. He is also a professional magician and frequently performs at the Magic Castle in Hollywood. He has demonstrated and explained his calculating talents to audiences all over the world. In 2005, *Reader's Digest* called him "America's Best Math Whiz."

DR. MICHAEL SHERMER is a contributing editor to and monthly columnist for *Scientific American,* the publisher of *Skeptic* magazine (www.skeptic.com), the executive director of the Skeptics Society, and the host of the Caltech public science lecture series. He is the author of numerous science books, including *Why People Believe Weird Things, How We Believe, The Science of Good and Evil, The Borderlands of Science,* and *Science Friction.*